CONTRIBUTORS

BILLARD, R.   *INRA, Hydrobiologie, 78 Jouy en Josas, France.*
BLAXTER, J.H.S.   *SMBA, Dunstaffnage, Oban, Scotland.*
BRAITHWAITE, H.   *Vickers Oceanics Ltd., PO Box 8, Barrow-in-Furness, England.*
BRETON, B.   *INRA, Hydrobiologie, 78 Jouy en Josas, France.*
ERIKSSON, L-O.   *Department of Ecological Zoology, University of Umeå, Umeå, Sweden.*
GIBSON, R.N.   *SMBA, Dunstaffnage, Oban, Scotland.*
GIRSA, I.I.   *A N Severtsov Institute of Evolutionary Morphology and Ecology of Animals, Leninskii Prospekt 33, 117071, Moscow V-71 USSR.*
GROOT de, S.J.   *Netherlands Institute for Fishery Investigation, Ijmuiden, Netherlands.*
MANTEIFEL, B.P.   *A N Severtsov Institute of Evolutionary Morphology and Ecology of Animals, Leninskii Prospekt 33, 117071 Moscow V-71, USSR.*
MATTY, A.J.   *Department of Biological Sciences, University of Aston, Birmingham, England.*
MITSON, R.B.   *MAFF Fisheries Laboratory, Lowestoft, Suffolk, England.*
MÜLLER, K.   *Department of Ecological Zoology, University of Umeå, Umeå, Sweden.*
MUNTZ, W.R.A.   *Department of Biology, University of Stirling, Stirling, Scotland.*
OLLA, B.L.   *NOAA/NMFS, Northeast Fisheries Center, Sandy Hook Laboratory, Highlands, New Jersey 07732, USA.*
PAVLOV, D.S.   *A N Severtsov Institute of Evolutionary Morphology and Ecology of Animals, Leninskii Prospekt 33, 11707 Moscow V-71 USSR.*
PRIEDE, I.G.   *Department of Zoology, University of Aberdeen, Aberdeen, Scotland.*
SCHWASSMANN, H.O.   *Department of Zoology, University of Florida, Gainesville, Florida, 32611 USA.*
SIMPSON, D.   *NSHEB Research Laboratory, Pitlochry, Perthshire, Scotland.*
SIMPSON, T.H. *DAFS, Marine Laboratory, PO Box 101, Aberdeen, Scotland.*
STAPLES, D.J.   *CSIRO, Northeastern Regional Laboratory, PO Box 120 Cleveland, Queensland, 4163, Australia.*
STUDHOLME, A.L.   *NOAA/NMFS, Northeast Fisheries Center, Sandy Hook Laboratory, Highlands, New Jersey 07732, USA.*
WAINWRIGHT, A.W.   *Centre for Research on Perception and Cognition, University of Sussex, Brighton, England.*

Rhythmic Activity
of Fishes

Rhythmic Activity of Fishes

Edited by

J. E. THORPE

Freshwater Fisheries Laboratory
Department of Agriculture and Fisheries
Faskally, Pitlochry
Perthshire, UK

1978

ACADEMIC PRESS

London . New York . San Francisco

A Subsidiary of Harcourt Brace Jovanovich, Publishers

ACADEMIC PRESS INC. (LONDON) LTD.
24/28 Oval Road,
London NW1

United States Edition published by
ACADEMIC PRESS INC.
111 Fifth Avenue
New York, New York 10003

Library of Congress Catalog Card Number: 78-52096
ISBN: 0-12-690650-5

Printed in Great Britain by
Galliard (Printers) Ltd, Great Yarmouth

PREFACE

Animals are confronted with a standing problem: the environment changes continually in a regular oscillatory manner, those oscillations determined by the elliptical path of the earth's annual revolution around the sun, its daily axial revolution and the moon's daily revolution about the earth. Along with this constant rhythmicity there is cyclic variation in light intensity and temperature, but these latter variables may be modified by local random events. For example, in a river, melting snows may disrupt the smooth increase of temperature patterns in spring, and floods reduce the light intensity through increased loads of suspended material. It has been demonstrated repeatedly that living organisms respond to this oscillating environment in a rhythmic way. The perception of rates of change of physical variables such as light intensity, or the exposure to total quantities of solar energy, may initiate a train of physiological events which serve to prepare the organism to meet the biological consequences of a changing environment. This adaptive function is itself achieved through rhythmic endocrine control.

Although much of the evidence for cyclically repetitive behaviour patterns and their control mechanisms in animals has been obtained from terrestrial forms, information on rhythmicity of activity in fishes is accruing rapidly. As this information is fragmented over a wide spread of technical journals it seemed desirable to draw it together, and so a symposium was organised by the Fisheries Society of the British Isles with this as its main objective. The meeting was designed around four main topics: the endocrine basis of rhythmic behaviour; physiological and behavioural rhythms; temporal aspects of community structure; and methods and instrumentation for use in the investigation of problems in these areas. Sixteen experts were invited to contribute review papers, and their contributions form the substance of this book. In addition, offers of short papers on current research topics were called for, and the thirty seven that were read at the meeting are noted on page 289. We hope that many of these will appear ultimately as papers in the *Journal of Fish Biology*.

The meeting was held at the University of Stirling, from July 4 - 8th 1977, and our thanks are due to Mr. J. Riddy for his efficient and orderly arrangement of facilities and accommodation for us. I should also like to thank the members of the Symposium steering committee, namely, Drs. J.H.S. Blaxter, R.N. Gibson, A.J. Matty,

R.I.G. Morgan, T.H. Simpson, P. Tytler and Messrs. R. Lloyd, R. Mitson and A.H. Young, for their help in the planning and organising of the meeting.

Professor Hans Meidner welcomed the participants at the opening session, and subsequent sessions were chaired by Professor W.R.A. Muntz, Professor H.O. Schwassmann, Dr. T.H. Simpson, Dr. S.J. de Groot, Dr. L-O Eriksson, Mr. E.D. le Cren, Dr. W.C. Clarke and Dr. G.P. Arnold: to all of these, and to the contributors, we are deeply indebted. I also acknowledge the valuable editorial help I have received from Drs. Gibson, Simpson and Tytler. Finally, it is a pleasure to record my thanks to the local secretary, Dr. Tytler, and to my colleague Dr. Morgan, for ensuring the smooth running of the whole enterprise in ways too many to catalogue.

JOHN THORPE
Pitlochry
September 1977

CONTENTS

LOCOMOTOR ACTIVITY OF FISH AND ENVIRONMENTAL OSCILLATIONS

KARL MÜLLER*

*Department of Ecological Zoology,
University of Umeå,
S-901 87 Umeå,
Sweden.*

Introduction

Rhythmic behaviour is one of the basic properties of living systems. The spectrum of biological rhythms includes periods with a duration of a second only, rhythms reflecting the 24 h period, and the annual cycle, as well as cycles that last over several years. Within this spectrum two types of rhythms may be distinguished: oscillations whose continuation is dependent upon rhythmic impulses received from the environment, and self-sustained oscillations. The pulsation of the heart provides a classic example of the latter type.

From an ecological point of view I shall not discuss the theoretical problems around the endogenous, circadian rhythmicity. It has been demonstrated by many workers over the last 4 decades, that the diurnal rhythms are primarily endogenous, i.e. that they are inherent characteristics of the living system. The experimental evidence for this conclusion is well established for several insects, birds and mammals, but may be less known in all organisms living in running water ecosystems.

Since the development of adequate methods, fish rhythmicity has recently been studied in many parts of the world. Our investigations are concentrated on a small river, Kaltisjokk, in northern Sweden, situated near the Arctic Circle ($66^0 42'N$, $20^0 25'E$). The extreme conditions in the subarctic areas, with extremely long days in the summer and short days in the winter, offer an excellent investigation area for biorhythmic research, for the reac-

---

*Footnote: Supported by the Swedish Natural Research Council, the Max-Planck-Gesellschaft and the Deutsche Forschungsgemeinschaft.

tion of the circadian system to the 'zeitgeber' conditions
(synchronizer, time-giver, time-signal) prevailing in the
course of the year. The life processes are concentrated
to the short summer period (3-4 month). During seven to
eight months of the year the river Kaltisjokk is covered
by ice.

These environmental conditions nevertheless render the
subarctic region an ideal natural experimental area for
studying ecological problems of Biorhythmics.

General Remarks

The 24 h cycle and the constantly changing light-dark
period causes variations in the abiotic factors in a running water-system, either directly or indirectly.

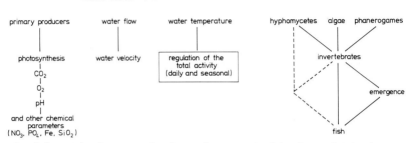

Fig. 1. Schematical general view of several abiotic and biotic oscillations in a running water.

Fig. 1 gives a schematic view of the different abiotic
and biotic oscillations. Photosynthesis causes daily variations in several abiotic factors, e.g. oxygen, carbon
dioxide and pH. Other metabolic processes, e.g. those of
algae, cause variations in the concentration of nitrate,
phosphate, iron and silicon.

The waterflow in a running water undergoes typical
seasonal variations which are specific for the different
geographical areas. Water temperature is directly affected by solar insolation, following a daily cycle with a
maximum in the early afternoon. The biotic processes in
running water plants and animals thus show a distinct temporal organization given by the rotation of the earth.

The fact that many factors in running water oscillate
with a 24 h period raises the question as to which of
these factors constitute the 'zeitgebers' for the diel
periodicity of stream organisms. The water temperature
in the River Kaltisjokk is constant within 0.4°C from
November to May, and all chemical factors are more or
less stable, the oxygen saturation being nearly 100%.

The invertebrates and the fish always begin and end
locomotor activity around the 5 lx threshold (Fig. 2).

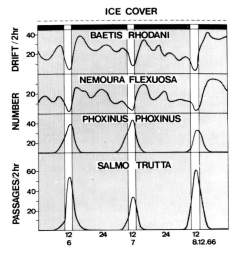

Fig. 2. The locomotor activity patterns of two freshwater insect larvae (*Baetis rhodani* (Ephemeroptera) and *Nemoura flexuosa* (Plecoptera)) and two fish species: Minnow (*Phoxinus phoxinus*) and Brown trout (*Salmo trutta*) in wintertime with constant water temperature (0.4°C). The activity of the larvae is measured by the drift in the River Kaltisjokk under a cover of ice and snow (1.20 m). The fish activity is monitored by photocell passages in time in artificial tanks, but with continuous water inlet from the river.

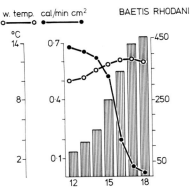

Fig. 3. Diel migrations of *Baetis rhodani* (Ephemeroptera) in relation to varying light intensities.

Although the water temperature in the River Kaltisjokk is constant, the larvae of *Baetis rhodani* (Ephemeroptera), of *Nemoura flexuosa* (Plecoptera), the brown trout (*Salmo trutta*) and the minnow (*Phoxinus phoxinus*) show a distinct change between activity and rest. The water temperature variations therefore cannot function as a 'zeitgeber' in

4 K. MÜLLER

wintertime.

The above mentioned chemical factors are even less directly influenced by insolation. Abundance and physiological conditions of stream algae have a considerable effect on the extent of variations of these factors, indicating that they are unsuitable as 'zeitgebers'. Only one factor shows consistent variations, the daily light-dark changes.

The importance of daily light-dark changes for locomotory activity can be demonstrated by an experiment: the locomotion of larvae of *Baetis rhodani* from the underside of a stone to the upper side is observed in varying light conditions (Fig. 3). Decreasing light intensity results in increasing activity and changing of the habitat of these larvae.

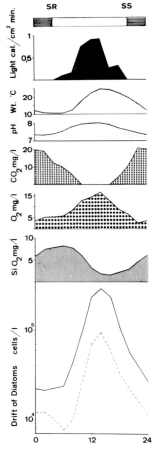

Fig. 4. The diel fluctuations of light, water temperature, pH-value, carbon dioxide, oxygen, silicon and the daily drift of two diatom species in a saline creek (51°N, West Germany).

Abiotic Oscillations

The interrelation between abiotic and biotic factors was clearly seen in our investigations in Central Europe. The extent to which periodic processes affect the chemical factors in a stream has been demonstrated by Müller-Haeckel (1965, 1966). The diel variations of oxygen, carbon-dioxide and pH were affected by the course of photosynthesis and the variation of silicon was due to the cell division of the diatoms (Fig. 4). Wolf *et al.* (1971) demonstrated the oscillations of CO_2 - concentration in a spring and at various distances along the stream. The amplitude of daily variations decreased at higher latitudes. But at the Arctic Circle the daily variations can still be about 1-2 mg/l dissolved oxygen if there is a large growth of green-algae.

Biotic Oscillations

The alternation of daylight and darkness provides a link between the organisms living in running-waters and the wider environment, through which environmental cues are obtained by the organisms. As in most other ecosystems, survival is dependent on a more or less exact timing of the various trophic functions: activities of the primary producers, the invertebrates and the fish fauna. All activity is in fact subject to chronobiological regulating mechanisms.

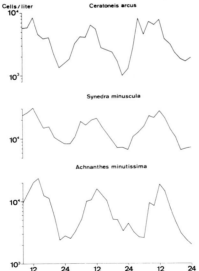

Fig. 5. Diel periodicity in the drift of three diatom species in the River Kaltisjokk in the beginning of June 1968.

Oscillations in Running Water Algae

Unicellular algae are important primary producers in

running waters. We have found that diatoms and other unicellular algae become detached from the substrate in a
distinctly rhythmic manner. Fig. 5 shows the diurnal rhythms of three typical diatoms in the River Kaltisjokk and
Fig. 6 the rhythm of a dominant green alga.

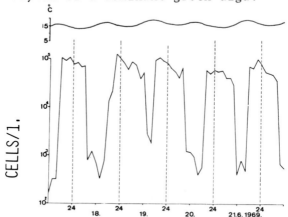

Fig. 6. Diel periodicity in the drift of a green alga (*Monoraphidium
Dybowskii*) in the River Kaltisjokk.

These diel cycles were studied by the drift of the algae.
The drift maximum of the diatoms, between 0800 and 1200 h,
coincides with the maximum food uptake of typical passive
feeders as well as of *Simulium*, Chironomidae and caddisfly
larvae. The coincidence of diatom drift and food uptake
of these insects holds true for the annual as well as for
the 24 h period.

    In the annual patterns of algae in running waters, the
situation in the autumn is most interesting. Ice forms in
the beginning of October in the Kaltisjokk. At this time
of the year diatoms occur in largest numbers. Diatoms and
the production of conidia by Hyphomycetes are also at a
maximum at this time. Insect larvae of the 'hiemal growth
type' (Brinck, 1949) have their highest growth rate, their
most frequent ecdysis, and their highest locomotor activity at this season (examples are: *Baetis rhodani, B. macani, Nemoura flexuosa, Capnia atra, Leuctra hippopus,
Diuva nauseni, Philopotamus montanus*).

The Oscillations of the Invertebrate Fauna

    Several authors (e.g. Tanaka, 1960; Waters, 1962;
Müller, 1963; Levanidova and Levandov, 1965; Elliott,
1965), working in different parts of the world, have shown
that the daily periodicity of the drift is a reflection
of the rhythmic nature of the animals' locomotor activity.
The number of organisms in the organic drift as a function

of time is an important variable in studies of diel and
seasonal periodicities in running water habitats, and
methods have been developed which permit the drift to be

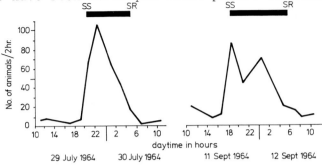

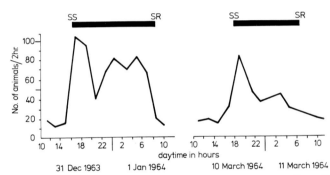

Fig. 7. Typical diel activity patterns of *Gammarus pulex* (Amphipoda)
measured in the drift in summer, autumn, winter and spring. SS = Sunset, SR = Sunrise.

automatically determined at regular intervals throughout
the year (Müller, 1965). The invertebrates in running-
water have species-specific activity patterns in the
course of a year. This pattern often includes a major
peak at the beginning of the active period and a minor
one towards its end, e.g. *Gammarus pulex*. We have found
that the variation in the basic activity pattern of this
amphipod is dependent on season or the length of the daily
photoperiod (Fig. 7).
    Fig. 8 shows a similar development of the rhythmic
activity pattern during an annual cycle for the drift of
Baetis (Ephemeroptera) in a woodland stream in northern
Sweden. In the short-day period (July, August), and in
spring (April, May) the activity pattern has a single
peak. In September-October and February-March the drift
shows a clear two-peaked 'bigeminus' configuration, and
in the long winter nights there is a three-peaked activity

pattern.
      The following experiments have been performed in both
Central Europe and in Swedish Lapland. Stretches of two

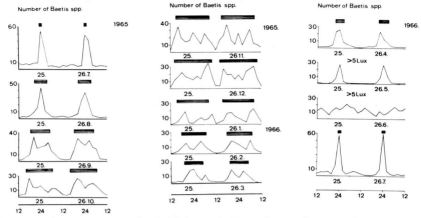

Fig. 8. The locomotor (= drift) activity of *Baetis* spp. (Number of
drifting larvae/2 h) on two consecutive days in every month during
one annual cycle in the River Kaltisjokk.

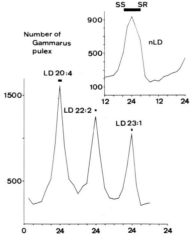

Fig. 9. Comparative studies on the locomotor activity (number of
drifting *Gammarus pulex*/2 h) in natural (nLD) and artificial light-
dark cycles (LD) (June 1964, Breitenbach, 51°N West Germany).

streams, one in each locality, were covered with sheets
of plastic. Under the covers we installed lamps that could
produce different light intensities and photoperiods.
Figs. 9-12 show that it was possible to modify the activity patterns of the following animals: (*a*) *Gammarus pulex*,
(*b*) a day-active caddis fly *Agapetus fusicpes* and (*c*)

Baetis rhodani. In the German stream we shortened or varied the length of the dark period.

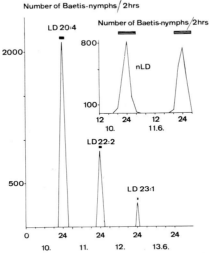

Fig. 10. Comparative studies on the locomotor activity (number of drifting *Baetis* larvae/2 h) in natural (nLD) and artificial light-dark cycle (LD) June 1964, Breitenbach, 51°N, West Germany).

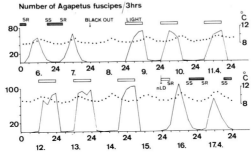

Fig. 11. The locomotor activity of the day-active larvae of *Agapetus fuscipes* (Trichoptera) in natural (nLD) and artificial light-dark cycles (LD) (April 1964, 51°N, Breitenbach, West Germany). SR = Sunrise, SS = Sunset).

In the Lapland stream we created a longer night period during the summer. In all these cases, the nymphs followed the artificial light-dark cycle. The results of these experiments clearly demonstrate that drift patterns are dependent on light-dark changes in both natural and artificial conditions.

The drift of limnic invertebrates is undoubtedly important for all species of fish as a source of food. The sig-

nificance of the geographical situation for the synchronization of the activity of insect larvae by the light-dark
cycle has been studied in three places in Europe.

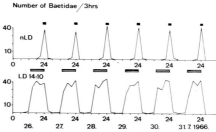

Fig. 12. Locomotor activity of larvae of *Baetis macani* and *B. subal-
pinus* (Ephemeroptera) in natural (nLD) and artificial light-dark
cycles (LD). (July 1966, Kaltisjokk, North Sweden).

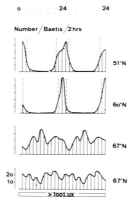

Fig. 13. The diel activity in larvae of *Baetis rhodani* (Ephemeroptera)
sampled at different latitudes in 2 h intervals between June, 20 and
June, 24 1963. $51^{0}N$ = Breitenbach, West Germany, $60^{0}N$ = Lindesberg,
Sweden, $67^{0}N$ = Kvikkjokk, Swedish Lapland.

The drift of the mayfly *Baetis rhodani* was simultaneously
measured in Central Europe ($51^{0}N$), in Central Sweden
($60^{0}N$) and in northern Sweden ($67^{0}N$). The results of these
investigations are summarized in Fig. 13.

It is seen that the pronounced daily periodicity occurring
at lower latitudes is absent in the subarctic summer.

Diel Oscillations in Emergence of Running-Water Insects

Unlike the drift the emergence of running-water insect
larvae is always synchronized with the 24 h period even
at higher latitudes. Many insects characteristically emerge
during the day, some at night, while others emerge at twilight. Emerging insects were continuously collected in the
River Kaltisjokk from May to October 1967 by means of an

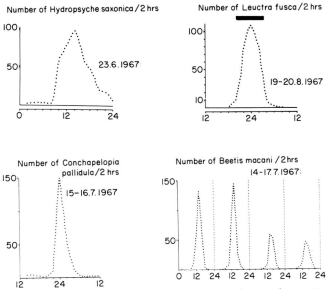

Fig. 14. The diel emergence activity of several running water insects in the River Kaltisjokk, Swedish Lapland: *Hydropsyche saxonica* (Trichoptera) *Leuctra fusca* (Plecoptera) *Conchapelopia pellidula* (Chironomidae) and *Baetis macani* (Ephemeroptera).

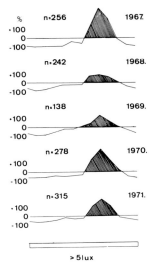

Fig. 15. The flight activity of the stonefly *Leuctra hippopus* (Plecoptera) 1967-1971 in the River Kaltisjokk. Two hour interval catches presented as percent deviation from the 24 h mean.

automatic sampler. The emergence pattern of four species
of Ephemeroptera, Plecoptera, Trichoptera and Chironomidae
are shown in Fig. 14. Similar patterns are found in both
hemi- and holometabolic insects.

Diel Oscillations in Flying Activity of Insects

   The diel flight periodicity in the number of imagines
of aquatic insects was studied by means of suction traps,
taking 2-hourly samples. The peak-times for flight occurred
in the afternoon and at dusk (Fig. 15).

Diel and Seasonal Periodicities in Fish

   The diel and seasonal patterns of fish activity that
are found in the River Kaltisjokk and at Messaure laboratory are typical for this latitude. Only two of the four
species (*Salmo trutta* and *Phoxinus phoxinus*) show the same
type of rhythmic patterns at other latitudes. The former
is crepuscular (activity peaks at dawn and dusk) and the
latter is day-active (Fig. 16).

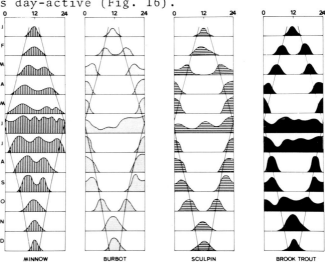

Fig. 16. Schematic general view of the diel activity patterns during
the year of the four fish species from the River Kaltisjokk: Minnow
(*Phoxinus phoxinus*), burbot (*Lota lota*), Sculpin (*Cottus poecilopus*)
and brown trout (*Salmo trutta*) after investigation from 1967 to 1975.

   The nocturnal species *Lota lota* and *Cottus poecilopus*
at high northern latitudes undergo a phase-inversion during the course of the year. In winter and until the first
half of February, both species are day-active with the
activity time starting at dawn. Around the middle of February, during a period of 10-15 days a phase shift to nocturnal activity occurs. Until April, the active period
coincides with the rapidly shortening night. When light

intensity no longer drops below 5 lx, the fishes respond
by extending their activity period to a maximum of 9 h.
At the time of continuous light the rhythm becomes dis-
sociated from the 'zeitgeber', the active period gradually
shifts from midnight to the afternoon, and sometimes activity is levelled out over the whole 24 h period.

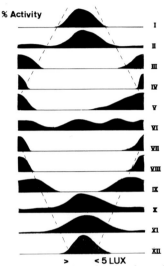

Fig. 17. Diel activity patterns in the burbot (*Lota lota*) with phase
inversions between night activity and day activity in the course of
the year at the Arctic Circle.

There is a tendency for free-running with circadian periods
always shorter than 24 h (= 22-23 h). In the second half
of July the burbot and the sculpin's activity is again
clearly entrained. Until the end of September both species
remain strictly nocturnal and attain a maximum activity
period of 10-11 h. Then the fishes shift again by 180^{0}
and become day-active during October (Fig. 17).
    The phase-inversion in these night-active species is
typical for high northern latitudes. It is possible to
induce phase-inversion artificially even at lower latitudes, by reducing the maximal light intensity to about
20-30 lx (Müller, 1978).
    We may conclude that the most important factor influen-
cing the diel activity patterns of fish is the alternation
between light and darkness. Water temperature is only of
secondary importance. Temperature determines the amount
of activity (Fig. 18), but not the phase-position (Fig. 19).
All chemical and other physical factors seem to be of
relatively minor importance.

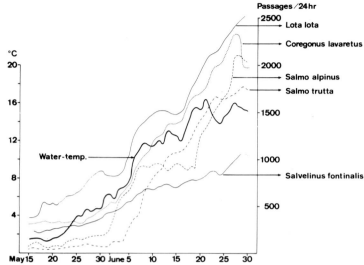

Fig. 18. The amounts of locomotor activity of fish species investigated at Messaure (Swedish Lapland) with increasing of water temperature in the early summer.

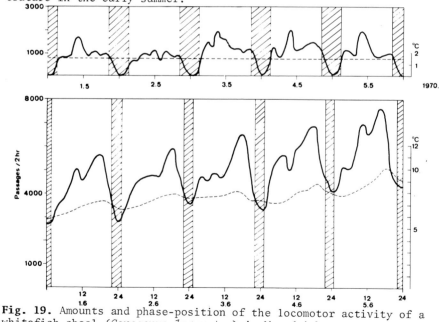

Fig. 19. Amounts and phase-position of the locomotor activity of a whitefish shoal (*Coregonus lavaretus*) in May (with constant water temperature) and June (with increasing water temperature).

Interactions Between Several Oscillations

The daily rotation of the earth around its axis has been a constant ecological factor throughout evolution to all organisms, including those inhabiting running water.

Every living organism survives by exploiting an ecological niche in its ecosystem. Often these niches also exist in time, the simplest expression of which is the division of organisms into diurnal, nocturnal and crepuscular. The natural consequence of this approach is to examine in more detail the function of timing mechanisms in the interactions of the various populations living together in a stream.

While concentrating on the temporal relations between organisms, we shall not, however, ignore the importance of all other dimensions that are encompassed by the concept of niche. But the feeding habits of fish in a stream, the availability and the possible selection of food are all functions of season and of the biotic oscillations in the course of the 24 h period.

Seasonal Interactions between Several Biotic Oscillations
Compared with streams at lower latitudes the River Kaltis- jokk represents a simple ecosystem which is characterized by a paucity of insect species, absence of amphipods and small biomass of molluscs. The emergence and flight of insects are strongly concentrated in to the short subarc- tic summer. Our investigations have been carried out in the lower part of the Kaltisjokk. Here the drift of dia- toms and that of the Baetidae (the dominant insect group), the total drift, the emergence, and the activity patterns of *Salmo trutta* have been continuously recorded (Fig. 20).

During most times of the year the drift of diatoms cor- responds with the size of their populations, with two yearly maxima, one in June-July and the other in September- November. A very close correspondence was found between the drift of diatoms and the drift of Baetidae which showed a maximum before emergence (June-July) during the maximum of growth of the *Baetis* nymphs (September-November). The positive correlation between drift of diatoms and Baetidae reflects the situation that the *Baetis* nymphs, which feed on diatoms are most active when the latter are most numer- ous.

A similar correlation between drift and food supply was found in the drift of all insects; the high number of drif- ting organisms during July-September consists of simulids and stoneflies (*Leuctra fusca, Amphinemura standfussi* emerging in September-October).

The surface drift, which reflects the number of emerg- ing insects is most intense in summer. 75% of about 200 determined water insects emerged during June-August in the Kaltisjokk.

The activity level of the poikilothermic *Salmo trutta*

K. MÜLLER

is primarily a function of the water temperature, which
to a high degree influences food intake and food diges-
tion as well.

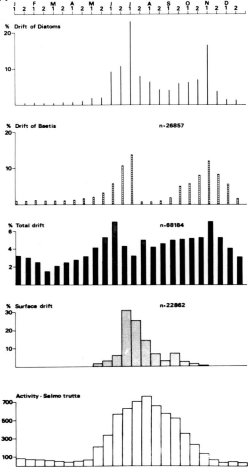

Fig. 20. Percentage of diatoms and invertebrates in the drift compared with the locomotor activity of *Salmo trutta* during one annual
period (after investigations in the River Kaltisjokk 1965-1974).

The short subarctic summer provides optimal temperature
conditions for the brown trout (monthly mean 12-14^{0}C)
combined with concentration of all food resources.
    These food resources include all drifting organisms,
emerging and emerged (flying) insects, ovipositing insects,
and, to a large extent terrestrial insects floating on the
water surface. The food uptake clearly reflects the seasonal variations in the occurrence of the several metamorphic

stages of the stream's insect fauna, and the decrease of emerging water insects in August results in a greater exploitation of terrestrial food components. This seasonal adaption to the available food supply has been well expressed by Hynes (1972): 'The fish are opportunists and eat what is available at the time'.

Abiotic and biotic oscillations throughout the course of the year in this example are an illustration of the temporal aspects of the trophic relations in an ecosystem.

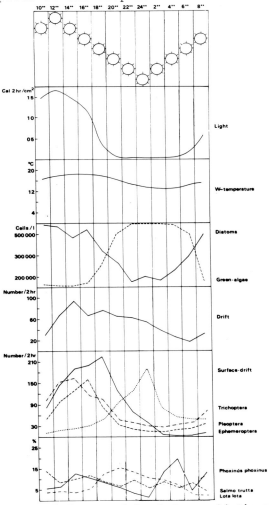

Fig. 21. Comparison of phase-position between abiotic and biotic oscillations at midsummer in the River Kaltisjokk.

Daily Interactions between Abiotic and Biotic Oscillations
I shall exemplify the interactions between predator and
prey in two examples. The first occurs in midsummer, when
most biotic processes are arhythmic or not synchronized
with the 24 h period, and the second under synchronized
conditions with a clear alternation between night and day.
In field observations and in laboratory experiments we
have found that insect nymphs, and fish in the subarctic
summer are not synchronized with the 24 h period. In contrast to this, emergence and flight of insects are always
synchronized. From about 15 June to 15 July in the River
Kaltisjokk, four weeks after the break-up of the ice,
there is a high level of metabolic activity among all
organisms from the algae to the fishes. Fig. 21 depicts
the functioning of an unsynchronized or partly synchronized system. The food supply at all levels of the food
chain is maximal throughout the day.
    Aperiodicity, especially among fish in midsummer is
probably of selective advantage. The subarctic summer
period of organic growth is short, and it is necessary
to exploit fully the opportunities for growth. A bloom
of algae represents a superabundant supply for aquatic
insect larvae. Feeding activity of these larvae, as reflected by their occurrence in the drift, is accordingly
continuous or aperiodic, as is the locomotor activity of
fishes which feed on the insects. On the other hand the
adult insects are confronted with greater diel fluctuations in air temperature and humidity, and for that reason a pronounced daily rhythmicity is retained.

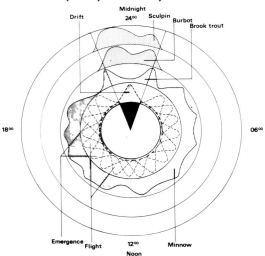

Fig. 22. Comparison between several biotic oscillations at the
beginning of August in the River Kaltisjokk.

In the second example (Fig. 22) there is a clear alternation between night and day to which the organisms must adjust. The main food items of the four species of fish (brown trout, minnow, burbot and sculpin) are drifting, emerging and emerged insects. The activity of the nocturnal and crepuscular fishes (*Lota lota*, *Cottus poecilopus* and *Salmo trutta*) show a close correlation with the activity of the drift and these species supplement this food with large numbers of regularly and irregularly occurring floating food items, such as surface-drifting aquatic and terrestrial insects. The day-active minnows on the other hand, were found to feed on day-emerging mayflies and chironomids, as revealed by the analysis of stomach contents.

Conclusion

In the functioning of the whole system of running-water, diel and seasonal rhythmicity play an essential part in the overall adjustment and survival of organisms. Indeed, all parts in the ecosystem are structured with regard to the temporal as well as the spatial dimensions.
The interactions between abiotic and biotic oscillations and interspecific oscillations are basic properties of the ecological organization of a stream.
The analysis of 'time-niches' in a running water are in an initial stage in the analysis of the function of a running water. They are a necessary complement to the studies of 'space-niches' in a stream.

References

Brinck, P. (1949). *Opusc. Ent. Suppl.* **11**, 1-250.
Elliott, J.M. (1965). *Nature* 205, 1127-1129.
Hynes, H.B.N. (1972). *The Ecology of Running Waters*. Liverpool Univ. Press 555 pp.
Levanidova, I.M. and Levanidov, V.Ya. (1965). *Zool. Zh.* **44**, 373-385.
Müller, K. (1963). *Nature* 198, 806-807.
Müller, K. (1965). *Limnol. Oceanogr.* **10**, 483-485.
Müller, K. (1978). The flexibility of the circadian system of fish at different latitudes. This volume, pp.
Müller-Haeckel, A. (1965). *Oikos* **16**, 232-233.
Müller-Haeckel, A. (1966). *Hydrobiologia* 28, 78-87.
Tanaka, H. (1960). *Bull. Freshwater Fish. Res. Lab. Tokyo* 9, 13-26.
Waters, T.F. (1962). *Ecology* **43**, 316-320.
Wolf, E.G., Cushing, C.E. and Rabe, F.W. (1971). *Limnol. Oceanogr.* **16**, 577-580.

PINEAL AND SOME PITUITARY HORMONE
RHYTHMS IN FISH

A.J. MATTY

*Department of Biological Sciences,
University of Aston in Birmingham.*

Pineal

This is a short review of the way in which the pineal
and the pituitary gland hormonal activity may control,
modulate, be controlled or otherwise be involved in the
rhythmic behaviour of fishes. Until comparatively recently
there was much discussion on whether or not the pineal
should be regarded as an endocrine organ. Research on
this structure and reporting on its activity seems now
to be fully the domain of the endocrinologist and comparative endocrinologist. In fact, to quote one of the world's
foremost contemporary authorities on pineal research:
'Among scientists, if not among fiction writers, the
pineal gland has finally come out of the realm of the
occult. The pineal is now known to be an inextricable
part of that closely-knit congeries of servomechanisms
referred to as the endocrine system. Anyone who would
deny this would be in error.' (Reiter, 1973.) It is not
proposed to discuss the role of the pineal as the seat
of the soul of fish or for that matter to look at the
pineal in comparative terms of a vertebrate. This has
been well reviewed by others (Hafeez, 1971). I rather
wish to look briefly at the investigations that have
been made in the last few years on pineal function in the
light of similar observations that have been made with
mammals.

Young (1935) showed that in the ammocoete larvae of
the lamprey removal of the pineal complex caused inter-
ruption of the daily rhythm of the integumentary melanophore which remained after the operation in an expanded
state under all levels of illumination. It was thought
at this time that the paling of the ammocoete when it
moved from light to darkness resulted from an inhibition
of pituitary melanophore hormone as a result of nervous
impulses set up by change of illumination of the pineal

complex. One of the earliest concepts of pineal function
was that it was the third eye of the lower vertebrates.
The more recent work on both *Lampetra planeri* and two
other species, *Geoteri* and *Mordacia*, have indicated that
the response of the pineal to light in cyclostomes might
be both photosensory and also endocrinological (Eddy and
Strahan, 1968; Meiniel and Collin, 1971; Joss, 1973.)
Pinealectomy abolishes the colour change rhythm and a substance has been identified in the gland (probably melatonin) which causes paling.

Most pineal research in the past decade has been on
the bony fish, but a small amount of work has been carried
out on elasmobranchs. Rüdeberg (1969) has demonstrated a
photoreceptor capacity of the pineal of the dogfish *Scyliorhynus caniculus* and also quoted evidence of light responsiveness, with spontaneous activity of the pineal
nerve fibres being inhibited by light, this duration of
inhibition being proportional to light intensity. The
effects of removal of the pineal have not been studied in
any elasmobranch, but in our own laboratory we have looked
at the effect of light on the pineal of *Scyliorhynus* and
the effect of continuous light and dark regimes have been
examined, and it has been shown that morphological changes
occur within the various pineal cells with the absence of
light. However, continuous illumination produced no struc-
tural effect within the cells. No diurnal changes however
could be observed in the histology of the cells, neither
could melatonin be isolated (Mechan, 1971).

It was the discovery in mammals of this substance, 5-
methoxy-*N*-acetyl-tryptamine, or melatonin, that has brought
about in the last decade much active experimentation in
fish pineal physiology. Evidence up to date suggests a
physiological role for the pineal, at least in some species, in relation to photoperiodic phenomena, both on a
daily basis in terms of the fishes melanophore response,
and on a seasonal basis in terms of the gonadotrophic responses, (the pituitary gonadal stimulating hormone).
The removal of the pineal gland has been carried out in
a number of fish species to investigate other possible
roles, but results have been somewhat equivocal. This is
possibly due to the fact that, in some species of fish,
the anterior brain region appears to be photoreceptive
in itself, and this makes the interpretation of such experiments difficult, and the effect of pineal removal
often seems to have been dependent on the involvement of
neighbouring brain regions (Ralph, 1975; Reiter, 1973).

Also, the time of year in which these experiments have
been performed seems to influence the results obtained
(Fenwick, 1970a, 1970b). Pang (1965) removed the pineals
of *Fundulus heteroclitus* and demonstrated that the pineal
was light-sensitive. However, more recent work by Hafeez
and Quay (1970a) has shown in the rainbow trout that al-

though both by its photosensory and endocrine capacities, the pineal in this species can have a contributory role in the mediation of responses to changes in environmental illumination, pinealectomy either with or without blinding, had no detectable effect on phototaxis. However, pinealectomy did abolish the day/night differences seen in blinded fish. The relative role played by a melatonin or by a nervous contribution is unknown. Whereas melatonin is active and specific as a paling agent in some species of fish, in others it can act as a darkening agent (Reed, 1968; Reed, *et al.*, 1969). In the pencil fish, the addition of melatonin to the spring water produces the night colouration within a few minutes depending on concentration, this being due to the fact that the melanophores of the night spots expand and the day band melanophores contract. This provided the first indirect evidence for the involvement of melatonin in the control of circadian pigment rhythms in fish.

Melatonin has now been identified in the pineal of fish (Fenwick, 1970a) but, due to the minute amount present, recent investigations have been concerned with observations on the enzyme which converts acetyl-seratonin into melatonin. Hafeez and Quay (1970b) have measured the presence of this enzyme in both the American roach and the rainbow trout. While there appeared to be more enzyme activity in the trout than the roach, in both these species the enzyme activity was found to be independent of constant light and constant darkness. However, more recent work by Smith and Weber (1974, 1976a, 1976b) has demonstrated a diurnal fluctuation in acetyl-seratonin-methyltransferase activity in the pineal gland of the steelhead trout and also alterations in levels of this enzyme under different light regimes. An interesting aspect of the work is that this pattern of activity of the enzyme could be abolished by blinding the fish, but not by surgical capping of the pineal region. Four days after capping a diurnal rhythm is still observed. These workers also found that unoperated fish held for six days in constant darkness failed to display an endogenous pineal ASMT-ase activity rhythm. Other workers have also shown that background colour plays an important role in the manifestation of the pineal ASMT-ase response to photoperiods in the steelhead trout (Smith and Weber, 1976b).

Having established a photoresponse, possibly hormonally mediated, and controlling both circadian behaviour and pigmentation in fish, we must now consider the role of the pineal in its relation to reproduction in fish. The pineal/gonadal relationship is well known in mammals, although the precise and particular mechanism and action of the pineal in reproduction is not yet obvious (Reiter, 1973). The most important apparent function of the pineal, certainly in some experimental animals, is its ability to

control or modulate seasonal reproductive rhythms. In the
hamster, under short photoperiods either in the laboratory
or under natural day lengths, the gonads become completely
degenerated, while pinealectomy prevents dark-induced gonadal involution. Thus, the conclusion has been reached in
these animals that, under conditions of restricted illumination such as would occur during the winter, a pineal
hormone effectively suppresses the reproductive states.
Whether the hormone in question in mammals is melatonin
itself or some related biogenic amine, or is a yet uniden-
tified polypeptide such as we see in the releasing hormones of the hypothalamus, is unknown. Melatonin itself is
not seen to be an effective anti-gonad agent in mammals.
Recently though, in the fish *Fundulus*, Vlaming and his co-
workers (Vlaming *et al.*, 1974) have shown that melatonin
treatment depending on the period in the year in which it
was administered, does affect the gonad development. Mela-
tonin-treated *Fundulus* collected in January or May and
maintained on the long photoperiod had a significantly
smaller gonad size than control animals, and this effect
was independent of the time of day that it was administered. Also, melatonin treatment of male fish maintained
on a short photoperiod in May retarded testis enlargement,
but similar treatment on animals kept on a short photoperiod in January had no effect. Many years ago, bull
pineal gland extract was fed to *Poecilia reticulata* and
secondary sexual characteristics were delayed, whereas
in the same species, pinealectomy accelerated sexual maturation (Krockert, 1936). Until recently, though, most
investigators have been of the opinion that pinealectomy
has little influence on gonadal activity in teleosts
(Schonherr, 1955; Rasquin, 1958; Pang, 1967; Peter, 1968).
Fenwick (1970b), working on the goldfish, has now shown
that gonadal size is seasonally pineal dependent in fish.
However, it is not known precisely the mechanism by which
melatonin inhibits gonadal function, or the nature of the
seasonal and photoperiod variation in response to melatonin. Whether or not melatonin itself releases some other
hormone, such as the polypeptide discovered in mammals,
is a matter for further experimentation. In a series of
experiments on the medaka (*Oryzias latipes*) Urasaki (1972a,
1972b and 1973) has shown that the presence of the pineal
is, at different seasons, required for oviposition and
for sustaining ovarian function during long day illumination, while the same organ inhibits the gonads under
short illumination conditions. Therefore the pineal may
not be simply a brake on reproduction, but may exert its
effects either by inhibition or stimulation, or by both
at different times. Without a doubt it seems that the
pineal of fish acts in some way as a dosiometer and transducer of radiations effecting reproductive and pigmentary
responses. There appears to be no experimental work on

the relationship of temperature or salinity changes to
pineal activity. However, it has been found that melatonin
decreases swimming activity, and that this hormone decreased locomotor activity occurs only during the light
phase of the photoperiodic cycle (Hafeez and Quay 1970a;
Byrne 1970). Another substance occurring in the brain of
the teleost fish (5-H-T - serotonin) increases locomotor
activity during the dark period. Circadian rhythms of
brain serotonin have been correlated with swimming activity in *Fundulus*, high concentrations occurring at two
peaks during the period of 24 hours. In male *Fundulus*
kept in constant darkness for a week there was significantly less brain 5-H-T than in males kept in constant light
and it has been suggested from these experiments that
5-H-T has a normal role in regulating the swimming activity of this fish (Fingerman, 1976). The endogenous rhythm
of serotonin that has been established in the pineal of
mammals has not yet been clearly established for fish
(Snyder *et al.*, 1967). This also applies to the neuro-
transmitter noradrenalin. (The detailed model for mammals
that has been postulated by Axelrod (1974) for pineal
function holds that an increased discharge of noradrenalin at night stimulates the adrenergic receptors of the
sympathetic nervous terminals innervating the pineal,
which in turn cause an increase in the synthesis of the
serotonin-*N*-acetyltransferase inside the pineal cells,
and ultimately the synthesis of melatonin.) There is no
evidence for a biological clock being present in the
nuclei of the hypothalamus.

Although there have been morphological indications of
changes in the hypothalamic neuro-secretory system under
conditions of altered photoperiod, no biochemical experiments appear to have been made (Honma and Suzuki, 1968;
Sathyanesan, 1965). In the teleost *Porichthys notatus*,
after 15 days continuous light the staining intensity of
the cells of the preoptic nucleus diminished, and there
was a greater accumulation of AF$^+$ substance occurring
along the axonal pathway and in the infundibulum. Ayu
(*Plecoglossus altivelis*) when exposed to short day lengths
and long day lengths showed changes in the staining reaction of the nucleus lateralis tuberis. Some experiments
of our own carried out a number of years ago on the minnow (*Phoxinus phoxinus*) demonstrated that, when exposed
to continuous darkness, there was an increase in aldehyde
fuschin stainable material in the cells of the nucleus
preopticus and the nucleus supraopticus.

Pituitary

The adenohypophysis of the pituitary gland produces a
number of hormones including the gonadotrophins, the
thyrotrophic hormone, the adreno-corticotrophic hormone,
growth hormone and prolactin. The site of production of

most of these hormones in fish, particularly in the teleosts, has been identified. Numerous papers have shown that
there is both seasonal and diurnal variation in most cells
in the pituitary (basophils and acidophils) (Ball and
Baker, 1969; Schreibman *et al.*, 1973). Rhythms of pituitary
gonadotrophins of fish have been considered in another
paper of this Symposium (Billard and Breton, 1978).
   Prolactin is a hormone which, in mammals, is responsible
for the initiation of milk secretion and in birds for stimulation of the crop gland to result in crop milk, and in
at least one amphibian to initiate migration of the species
to water preparatory to mating and egg-laying. In fish,
prolactin has been associated with a number of activities
in different species and therefore it would not seem surprising that the hormone which has a multitude of general
metabolic actions and also actions associated with reproduction and behaviour would show cyclic and rhythmic vari-
ance.
   In the mid-sixties it was shown that injections of ovine
prolactin enabled hypophysectomized euryhaline fish to survive more easily in fresh water. (*Fundulus* - Pickford *et
al.*, 1965, 1966; *Poecilia* - Ball and Olivereau, 1964;
Xiphophorus - Schreibman and Kallman, 1966; *Tilapia* -
Handin *et al.*, 1964 and *Gambusia* - Chambolle, 1966). Lam
and Hoar (1966) continued these investigations by examining the seasonal effects of prolactin on fresh water osmoregulation of the stickleback (*Gasterosteus aculeatus*).
They injected prolactin into fish after the transfer of
them from sea-water to fresh water at different seasons
of the year in order to find out whether prolactin was
responsible for bringing about the physiological change
that enables the marine stickleback to tolerate fresh
water habit in spring and summer. They were able to show
that prolactin altered the fresh water osmoregulation of
the marine form of this stickleback at certain seasons.
During the past 10 years Ball and his colleagues (Batten
and Ball, 1976) have investigated the role of prolactin
in osmoregulation and iron balance. Although no change
in the histology of the prolactin cells has been seen
during the transition of fish from sea-water to fresh
(McKeown and Overbeeke, 1972), Batten and Ball (1976)
demonstrated circadian changes in prolactin cell activity
in *Poecilia latipinna* in fresh water. They were able to
show that intense synthetic activity in the cells occurs
during the period mid-day to evening compared with other
times during the period. Diurnal rhythm of pituitary prolactin activity was also demonstrated by Vlaming *et al.*,
(1975) who used a bioassay which depended on pigment dispersal in the xanthophores of another fish (*Gillichthys
mirabilis*). The specificity of this technique may be criticised, but Leatherland and his co-workers were able to
show at the same time, using an immunoassay technique,

that circadian rhythms occurred in the plasma prolactin of juvenile salmon (*Oncorhynchus nerka*) (Leatherland *et al.*, 1974). Both groups of workers correlated changes in prolactin with changes in plasma free fatty acids, which also show diurnal variations in level (Leatherland and McKeown, 1973), as have Spieler and Meier (1975). Recently, McKeown and Peter (1976) have investigated the effects of photoperiod and temperature on prolactin release in the goldfish pituitary gland where fish have been acclimated to different photoperiods and temperature and then serum and pituitary samples analysed by radio-immunoassay for prolactin. Longer photoperiods and higher temperatures caused pituitary prolactin release. Also, serum prolactin changed on a circadian rhythm, and the rhythm was modified depending on the length of the photoperiod. The maximum prolactin concentration in the blood occurs at the end of the scotophase while the minimum occurred in the middle to the end of the photophase. More information is needed to know if this is a direct effect or is consequent on osmoregulatory requirement or is due to some other secondary response.

Plasma growth hormone in fish has been shown to exhibit a circadian rhythm, peak values occurring in the mid-dark period (Leatherland *et al.*, 1974). The meaning of this in terms of the biology of the animal is not understood. The levels may reflect activity rhythms. The relationship between growth hormone and full fatty acid mobilization which occurs in mammals may also hold for fish.

Adrenocorticotrophin (ACTH), the pituitary stimulant of the adrenal hormones has not been assayed directly in fish pituitary or plasma in order to establish a possible seasonal/circadian rhythm. However, there have been a number of publications over the past decade which have indicated clearly diurnal rhythm in plasma corticosteroids and it would seem highly likely that when techniques are refined and the work done, then rhythm of ACTH will be seen to occur (Boehlke *et al.*, 1966; Srivastava and Meier, 1972).

Cyclic thyroidal activity has been reviewed in the past (Matty, 1960) and Simpson (1978) presented the current picture in this symposium. Rhythms in teleost thyroid stimulating hormone (TSH) almost certainly are present although no direct method of TSH measurement is available. One has to depend on measurement of thyroid activity to indicate TSH. Singh (1967) has indicated such a seasonal change in pituitary levels using a bioassay. In our laboratory Bromage and his co-workers have shown that the trout (*Salmo gairdneri*) thyroid responds to TSH injection by secretion of thyroid hormones. As several studies have suggested diurnal and seasonal changes in thyroid hormone it is likely that there are underlying changes of TSH. Whether these changes are ultimately produced by releasing

hormone produced by the hypothalamus is not known. This
applies to all the pituitary hormones. Mammalian thyroid
releasing hormone (TRH) *reduces* thyroid activity in the
trout and the inhibition is dose dependent and specific.
Luteinising hormone release hormone and melanocyte release
inhibiting hormone are without effect.

Having looked at the pineal and pituitary and rhythm
levels of such substances as melatonin, ASMT-ase, prolactin and growth hormone, what do we conclude? Firstly, it
would appear that they are all labile. They give no evidence of being biological clocks, i.e. show mechanisms
that time organismic rhythms that persist in constant conditions. There is little evidence for postulating either
an escapement clock or a non-escapement clock in the fish
pineal or pituitary gland. There is, however, evidence
that there is entrainment (that is the coupling of the
organismic rhythm to an external oscillation causing the
rhythm to display the frequency of the external oscillator).

In mammals biological clocks have been postulated to
exist in or near the supraoptic nucleus of the hypothalamus. Such a postulate might be made for fish, but it must
remain a postulate.

If one turns to the more simple analysis, i.e. that of
cause and effect, our biology is weak in understanding.
Do changes in free fatty acids induce growth hormone
changes in salmon plasma, or is it the other way round?
Are the rhythms of plasma prolactin in fish merely the
reflection of metabolic demands? I rather suspect that
they are.

Probably in fish the closest we have got to a hormonal
(or rather a tissue hormone) biological clock is in the
case of *Fundulus* where the nocturnal drop of serotonin in
the brain would appear to be a reflection of an endogenous
rhythm, but where prolonged exposure to darkness can also
work to reduce the level of brain serotonin.

Finally, the pineal complex in fish must make a photosensory contribution to, at least in trout, the day-night
coloration rhythm in blinded animals and to an endocrine
contribution in the regulation of this same rhythm. Pineal
modulation of fish reproductive rhythms will be, I suggest,
a subject of research for many years hence.

References

Axelrod, J. (1974). *Science*, 184, 1341-1348.
Ball, J.N. and Baker, B.I. (1969). *In: Fish Physiology* (W.S. Hoar
    and D.J. Randall, eds.), vol.II *The Endocrine System*, Academic
    Press, New York, pp.1-110.
Ball, J.N. and Olivereau, M. (1964) *Compt. Rend. Acad. Sci. Paris*,
    259, 1443-1446.
Batten, T.F.C. and Ball, J.N. (1976). *Cell. Tiss. Res.*, 165, 267-280.
Billard, R. and Breton, B. (1978). This volume, pp. 56-67.

Boehlke, K.W., Church, R.L., Tiemeier, O.W. and Eleftheriou, B.E. (1966). *Gen. Comp. Endocrinol.*, **7**, 18-21.
Byrne, J.E. (1970). *Canad. J. Zool.*, **48**, 1425-1427.
Chambolle, P. (1966). *Compt. Rend. Acad. Sci. Paris*, **262**, 1750-1753.
Eddy, J.M.P. and Strahan, R. (1968). *Gen. Comp. Endocrinol.*, **11**, 528-534.
Fenwick, J.C. (1970a). *Gen. Comp. Endocr.*, **14**, 86-97.
Fenwick, J.C. (1970b). *J. Endocr.*, **46**, 101-111.
Fingerman, Sue, W. (1976). *Comp. Biochem. Physiol.*, **54**C, 49-53.
Hafeez, M.A. and Quay, W.B. (1970a). *Z. vergl. Physiologie*, **68**, 403-416.
Hafeez, M.A. and Quay, W.B. (1970b). *Comp. & Gen. Pharmacol.*, **1**, 257-262.
Hafeez, Mohammad A. (1971). *J. Morph.*, **34**, 281-314.
Handin, R.I., Nandi, J. and Bern, H.A. (1964). *J. Exptl. Zool.*, **157**, 339-343.
Honma, Y. and Suzuki, A. (1968). *Jap. J. of Ichthyol.* 15, 11-27.
Joss, Jean M.P. (1973). *Gen. & Comp. Endocrinol.*, **21**, 188-195.
Krockert, G. (1936). *Z. Ges. Exptl. Med.*, **98**, 214-220.
Lam, T.J. and Hoar, W.S. (1966). *Canad. J. Zool.*, **45**, 509-516.
Leatherland, J.F. and McKeown, B.A. (1973). *J. Interdisc. Cycle Res.* **4**, 137-143.
Leatherland, J.F., McKeown, B.A. and John, T.M. (1974). *Comp. Biochem. Physiol.*, **47A**, 821-828.
Matty, A.J. (1960). *Symp. Zool. Soc. London*, **2**, 1-15.
McKeown, B.A. and Overbeeke, A.P. van. (1972). *J. Fish Res. Bd. Can.*, **29**, 303-309.
McKeown, B.A. and Peter, R.E. (1976). *Canad. J. Zool.*, **54**, 1960-1968.
Mechan, D.J. (1971). The pineal of some chondrichthyes. M.Sc. Thesis, University of Aston in Birmingham.
Meiniel, A. and Collin, J.P. (1971). *Z. Zellforsch. Mikrosk. Anat.*, 117, 354-380.
Pang, Peter K.T. (1965). *American Zoologist*, **5**, 254.
Pang, P. (1967). *American Zoologist*, 7, 715.
Peter, R.W. (1968). *Gen. Comp. Endocrinol.*, **10**, 443-449.
Pickford, G.E., Pang, P.K.T. and Sawyer, W.H. (1966). *Nature*, **209**, 1040-1041.
Pickford, G.E., Robertson, E.E. and Sawyer, W.H. (1965). *Gen. Comp. Endocrinol.*, **5**, 160-180.
Ralph, Charles L. (1975). *Amer. Zool.*, **15** (Suppl.1), 105-116.
Rasquin, P. (1958). *Bull. Amer. Mus. Natur. Hist.*, **115**, 1-68.
Reed, B.L. (1968). *Life Sci.*, **1**, 961-973.
Reed, B.L., Finnin, B.C. and Ruffin, N.E. (1969). *Life Sci.*, **8**, 113-120.
Reiter, Russel J. (1973). *Ann. Rev. Physiol.*, **35**, 305-328.
Rüdeberg, C. (1969). *Z. Zellforsch.*, **96**, 548-581.
Sathyanesan, A.G. (1965). *J. Morph.*, **117**, 25-48.
Schonherr, J. (1955). *Zool. Jahrb. Abt. Allgem. Zool. Physiol.*, **65**, 357-368.
Schreibman, M.P. and Kallman, K.D. (1966). *Gen. Comp. Endocrinol.*, **6**, 144-155.
Schreibman, M.P., Leatherland, J.F. and McKeown, B.A. (1973). *Amer.*

Zool., **13**, 719-742.
Simpson, T.H. (1978). This volume, pp.55-68
Singh, T.P. (1967). *Experientia,* **23**, 1016-1017.
Smith, J.R. and Weber, L.J. (1974). *Proc. Soc. Exptl. Biol. & Med.,* **147**, 441-443.
Smith, J.R. and Weber, L.J. (1976a). *Can. J. Zool.,* **54**, 1530-1534.
Smith, J.R. and Weber, L.J. (1976b). *Comp. Biochem. Physiol.* **53C**, 33-35.
Snyder, S.H., Axelrod, J. and Zweig, M. (1967). *J. Pharmac. exp. Ther.,* **158**, 206-213.
Spieler, R.E. and Meier, A.H. (1975). *J. Fish Res. Board Can.,* **33**, 183-186.
Srivastava, A.K. and Meier, A.H. (1972). *Science,* **177**, 185-187.
Urasaki, H. (1972a). *Annot. Zool. Jap.,* **45**, 10-15.
Urasaki, H. (1972b). *Annot. Zool. Jap.,* **45**, 152-158.
Urasaki, H. (1973). *J. Exp. Zool.,* **185**, 241-245.
Vlaming, V.L. de, Sage, M. and Charlton, Connie B. (1974). *Gen. Comp. Endocrinol.,* **22**, 433-438.
Vlaming, V.L. de, Sage, M. and Tiegs, R. (1975). *J. Fish Biol.,* **7**, 717-726.
Young, J.Z. (1935). *J. Exp. Biol.,* **12**, 254-270.

RHYTHMS OF REPRODUCTION IN TELEOST FISH

R. BILLARD and B. BRETON

I.N.R.A., Laboratoire de Physiologie des Poissons
78350 JOUY EN JOSAS, France

Introduction

Reproduction is a cyclic phenomenon which occurs periodically in fish, sometimes once every two or three years but more usually, once or several times a year. But if rhythm is defined as a succession of reproductive cycles, rhythmicity can hardly be an innate characteristic, since this cyclic reproductive activity is usually considered to result from adaptation to a changing environment mediated through the endocrine system. Fish live in a large variety of biotopes leading to a great diversity of reproductive patterns. This paper describes some types of reproductive cycle and their associated hormonal changes.

The reproductive cycle is divided into two parts, gametogenesis and spawning.

Gametogenesis is the formation of highly specialised gametes (oocytes or spermatozoa) from simple germ cells (spermatogonia or oogonia). The duration of gametogenesis varies according to species and rearing temperature. In *Oryzias latipes*, the preleptotene-early spermatid interval is 5 days at 25°C and 12 days at 15°C (Egami and Hyodo-Taguchi, 1967). The leptotene-spermatozoa interval is 12 days in *Oryzias* and 14 days in *Poecilia* (Billard, 1968). The total duration of spermatogenesis at 25°C was estimated at 36 days in *Poecilia* (Billard, 1969). In the viviparous species *Xiphophorus*, at 24°C, a reproductive cycle of 35 days includes a gestation period of 27 days (Siciliano, 1972).

Spawning comprises the sequences of events leading to the liberation of gametes, meiosis resumption, oocyte maturation, ovulation and oviposition in females and spermiation and sperm release in males. These processes are complex and more variable in duration.

The endocrine changes during the reproductive cycle include variations in the levels of pituitary and plasma

gonadotropin (GTH*) and plasma sex steroids. Much information on fish hormones is now available but heterogeneity
in the assays and the lack of international standards pre-
clude definitive interpretation. Methodologies for GTH
radioimmunoassay (RIA), GTH bio-assay and steroid analyses
have been discussed respectively by Crim *et al*., (1975),
Jalabert *et al*., (1974) and Sandor and Idler (1972).

Circannual Rhythm of Reproduction in Fish Living in Temperate and Cold Zones

Gametogenesis

Although gametogenesis and its endocrine correlates
have been described in only a few species of teleost fish,
some typical groups can be identified. Experimental variation of the environment gives some indication of the
factors determining the reproductive cycle.

Salmonids In this group, gametogenesis occurs in summer
and autumn and can be correlated with decreasing photoperiod and temperature.

In Europe, trout gametogenesis begins with vitellogenesis in females; the division of type B spermatogonia
in males starts in late spring or early summer. During
the summer, the gonadosomatic index (GSI) increases regularly and, normally, spermatogenesis ends in September
or October and vitellogenesis in October or November. An
example of a spermatogenetic cycle is given in Fig. 1.
A peak of t-GTH secretion is observed at the beginning
of the spermatogenetic cycle when a slight multiplication
of type A spermatogonia is seen in the testis and when
some type B spermatogonia appear. At the same time, high
levels of estradiol 17β are present in the plasma. During
intense testicular activity, leading to a sharp increase
of the GSI, there is a parallel rise of t-GTH and a slight
increase in the concentration of estradiol. t-GTH pituitary
content is maximum at the beginning and at the end of
spermatogenesis. A similar pattern of development was
found in studies of pituitary gonadotropic cells by immunofluorescence (Fig. 2). At the beginning of the oogenetic
cycle in female rainbow trout (Fig. 3), a high level of
estradiol and a slight rise in t-GTH were noted. t-GTH
pituitary content measured either by RIA or bioassay,
plasma t-GTH, estradiol, oocyte diameter and GSI rose
simultaneously.

Studies of reproductive development in Atlantic salmon
(chilko variety) and brook trout have been reported by

---

*Footnote: c-GTH : carp gonadotropin
            s-GTH : salmon gonadotropin
            t-GTH : trout gonadotropin

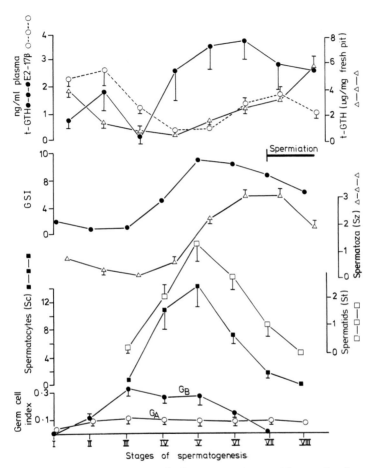

Fig. 1. Endocrine components of the spermatogenetic cycle in rainbow trout (Breton *et al.*, unpublished), RIA for GTH and estradiol 17β (E2), importance of the germ cell is measured by the germ cell index (Billard *et al.*, 1974).

Crim *et al.*, (1975). In salmon, plasma GTH rises slightly during vitellogenesis and attains very high levels during the spawning season. In brook trout males and females, as in rainbow trout, the rise of GSI is closely associated with a rise of t-GTH. Testosterone and 11-ketotestosterone levels rise before the spawning season in male brook trout (Sangalang and Freeman, 1974). Additional information on steroid hormones in the plasma of the Atlantic salmon is due to Schmidt and Idler (1962). They observed notable variations associated with migration and sexual maturation (see Ozon 1972 for review).

R.B. BILLARD ET AL.

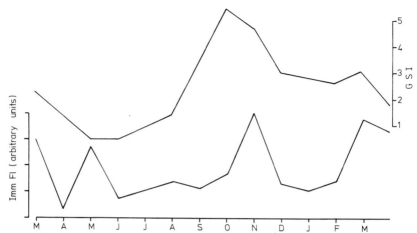

Fig. 2. Evolution of the fluorescence intensity of gonadotropic cells in the pituitary of male rainbow trout during the reproductive cycle; details of the immunofluorescence technique in Billard *et al.*, (1971); intensity and fluorescence is estimated by arbitrary units (from Escaffre *et al.*, unpublished).

Circannual rhythms have been investigated as a preliminary to attempts to advance the time of reproduction (see De Vlaming, 1974, for review). Recent work has confirmed that photoperiod variation is the main factor involved in gonadal development in salmonids (Breton and Billard, 1977). Constant long (16L-8D) or short (8L-16D) photoperiods did not induce gonadal recrudescence; only a decreasing photoperiod (16L-8D to 8L-16D) from February 24 to June 19 induced full spermatogenesis and spermiation. The amount of plasma t-GTH and the efficiency of spermatogenesis were significantly increased at 16° showing that temperature is a factor, though not limiting, in reproduction in trout.

Other species which show a similar annual reproductive rhythm are *Salmo salar* (Jones, 1940), *Salvelinus fontinalis* (Henderson, 1962), *Plecoglossus altivelis* (Honma and Tamura, 1962), *Embiotoca jacksoni* (Lagios, 1965), *Sebastodes paucispinis* (Moser, 1967) and *Cymatogaster aggregata* ♀ (Wiebe, 1968).

Cyprinids Gametogenesis usually starts in autumn, continues slowly during the winter and ends in the spring.

(a) *Roach* The reproductive cycle of the male roach has been described by Escaffre and Billard (1976). In

---

*Footnote: t-GTH : trout gonadotropin

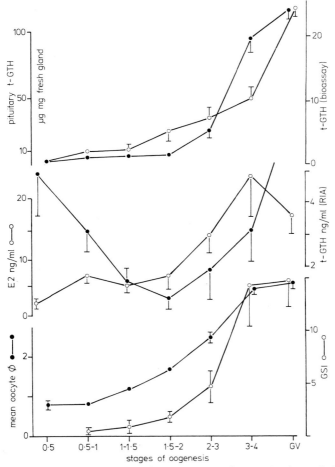

Fig. 3. Endocrine components of the oogenetic cycle in rainbow trout. Top graph: measurements of t-GTH content in the pituitary by RIA (•), and bioassay (o) (Breton *et al.*, unpublished).

autumn there is a rise of type B spermatogonia; meiosis starts in late December and is continuous until April. Spermiogenesis occurs in March, April and May and spermatozoa are very abundant in May and June. Pituitary concentrations of GTH exhibit some variation, rises occurring in autumn and late winter and in late spring at the time of spermiation.

(*b*) *Carp*   The pattern of the reproductive cycle in the carp is dependent on the thermal rearing regime. Some data are available on the c-GTH pituitary content in males measured by RIA (Fig. 4) and estimated by immunofluorescence (Fig. 5). Pituitary concentration of GTH (by RIA) decreases in the winter and rises during the spring at the time of gonadal recrudescence and the spawning season.

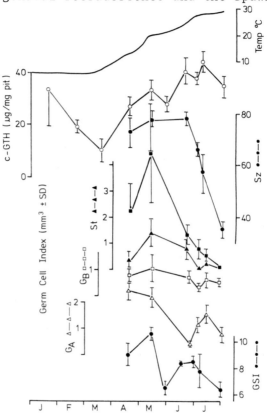

Fig. 4. Spermatogenesis in carp: evolution of the germ cells in the testis in spring and early summer. Variation of the pituitary content of immunoreactive c-GTH (µg/ml fresh gland) between January and July (From Weil, unpublished).

In another study (Fig. 5), spermatogenesis occured in spring and a second wave was observed in the summer. Gon-adotropic cell fluorescence was very strong in spring, during the first wave, and very weak in summer and early autumn during the second spermatogenetic wave. There may have been a cyclic depletion of the pituitary but this possibility requires further study. Eleftheriou *et al.*, (1968) have shown that the evolution of gonadal develop-ment is highly correlated with the level of plasma estra-diol 17β and to a much lesser extent with 16-ketoestradiol. Gonadal development is associated with rising temperature,

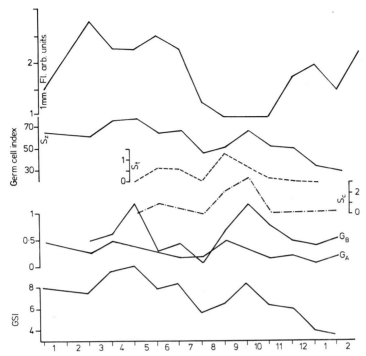

Fig. 5. Evolution of the GSI, spermatogenetic cycle and immunological fluorescence of GTH pituitary cells in carp *Cyprinus carpio* throughout the annual reproductive cycle (after Escaffre, unpublished).

and spawning occurs when the temperature cycle is at its maximum.

(c) *Goldfish*   In goldfish, c-GTH increases with gonadal recrudescence (Fig. 6). In the males, androgens also rise prior to the spawning season but estrogen does not (Schreck and Hopwood, 1974). In females, androgens and estrogens are present during the spawning season. The amount of pituitary c-GTH, estimated by immunofluorescence decreases during gonadal recrudescence but rises later, minimum fluorescence intensity occurring in summer as for carp. Other studies have been carried out on cyclic variation in goldfish. Beach (1959) found that the surface of the pituitary basophil cells first increases with increasing oocyte diameter and decreases near maturity. The neurosecretory material in the NPO also varies; it is very low in the spring, and high in summer and autumn (Mieszkowska and Jasinski, 1973).

(d) *Tench*   A peak of GTH concentration is observed in the plasma in February, at the beginning of the sperma-

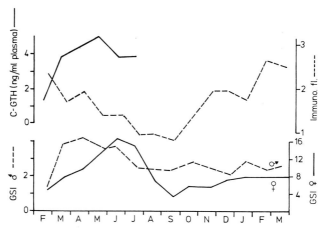

Fig. 6. Annual variation in GSI, intensity of fluorescence of the gonadotropic pituitary cells (males) and plasma c-GTH in goldfish (Escaffre and Gillet, unpublished).

togenetic cycle (Breton *et al.*, unpublished). It is low in April and rises subsequently as spermatogenesis progresses. It decreases slightly during the spawning season; pituitary concentration of GTH is very high in the middle of the spawning season. This experiment was carried out in Poland ($52°N$), where gametogenisis started in winter.

(e) *Other species* Other cyprinids e.g. *Couesius plumbeus* (Ahsan, 1966), *Abramis brama* (Shikhshabekov, 1974) and various other species such as *Hypseleotris galii* (Mackay, 1973), *Agonus cataphractus* (Le Gall, 1969), *Dicentrarchus labrax* (Barnabe, 1976) exhibit a reproductive cycle similar to that of roach and carp. Tench and other species have a different pattern of reproductive cycle, gametogenesis occurring entirely in rising photoperiod and temperature. Typical of these are, *Cymatogaster aggregata* male (Wiebe, 1968), *Pylodictus olivaris* (Turner and Summerfelt, 1971), *Paralabrax clatharus* (Smith and Young, 1966), *Notropis stramineus* (Summerfelt and Minckley, 1969), *Scomber scomber* (Bara, 1960), *Chromis chromis* (Contini and Donato, 1973), *Cynoscion regalis* (Merriner, 1975) and *Epinephelus* (Brusle and Brusle, 1975).

Pleuronectids In the female plaice, plasma steroids, such as cortisol, testosterone and estradiol are low in concentration in the summer, during the resting period and rise with gonadal recrudescence, (Wingfield and Grimm, 1977). Estradiol reaches a maximum and drops just before the peak spawning season, while testosterone and cortisol reach their maximum levels just at the beginning of the peak spawning period. Testosterone decreased during this period while cortisol levels fell later. GTH pituitary content was minimal prior to and during the spawning season. Except for estradiol, males exhibit similar seasonal changes. The highest steroid level was recorded in January and February before the peak spawning season. In immature plaice, a pronounced cortisol peak is observed, indicating that seasonal changes of this steroid are independent of the gonadal cycle and may be associated with feeding activity as suggested by Campbell *et al.*, (1976) for the winter flounder. In this latter species, seasonal variations in the peripheral steroid levels are closely associated with GSI, as in plaice: testosterone levels are higher in females than in males. 11-Ketotestosterone remains low throughout the year in females and appears to be more important in the males than testosterone (Campbell, 1975; Campbell *et al.*, 1976). Additional information on reproduction in pleuronectids is given by Pitt (1966) and Lahaye (1972).

Experimental modification of the reproductive cycle   In cyprinids, both temperature and photoperiod are important factors and may interact. In goldfish females (Fig. 7) taken in October, gonadal recrudescence is more rapid

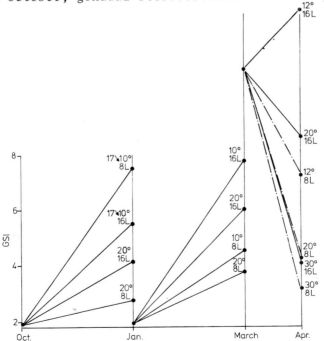

Fig. 7. Effects of various photoperiod-temperature combinations of gonadal development (GSI) in the goldfish (Gillet, unpublished).

when fish are under decreasing temperature (17→10°C) than
under constant elevated temperature (20°C). The effect of
photoperiod is dependent on the temperature regime. In
January, long day lengths are stimulatory especially when
associated with a low temperature (10°C). In March, when
oogenesis is more advanced, only long day and low temperature regimes are stimulatory. Experiments carried out
on tench, under three temperature regimes, show that GSI
peaks earlier and that the pattern of plasma GTH levels
is different in heated ponds (Fig. 8); the number of
oogenesis waves and consequently of spawnings is increased

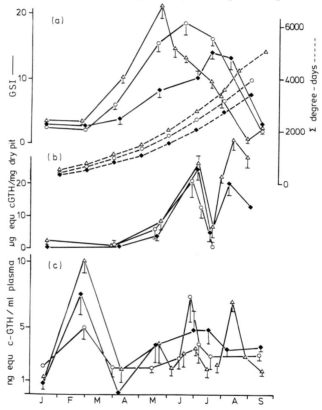

Fig. 8. Annual reproductive cycle of female tench under three thermal
regimes I (Natural), II (I + 3°C), and III (II + 3°C). A: GSI, B:
temperature (total degree days), C: pituitary content of immunoreactive gonadotropin (c-GTH equivalent), D: plasma gonadotropin (Breton
et al., unpublished).

at the higher temperatures (Breton *et al.*, unpublished).
GTH pituitary content is at its maximum during the inter-

val between successive spawnings.

Ovulation, Spermiation, Spawning

Spawning season Three main spawning seasons can be identified in France, late autumn and winter (salmonids),
early spring (pike, perch), and late spring and early
summer (tench, roach, gudgeon).
     Some species of fish, e.g. *Chromis chromis* (Contini
and Donato, 1973), *F. diaphanus* (Fournier and Magnin,
1975), *Notropis longirostris* (Heins and Clemmer, 1976)
reproduce in summer, but most of these are intermittent
spawners. The duration of the spawning period may depend
on the latitude; in the USSR for instance, the spawning
season of bream varies from 58-60 days in the north to
16 days in the Volga delta (Shikhshabekov, 1974). Spawning of the sea bass occurs in winter in the Mediterranean
area, in spring in Brittany and during early summer in
Ireland when the temperature reaches 10-12°C. The period
of running sexual products is usually longer than the
breeding season, especially for the males. Sperm release
begins before ovulation and can also be observed by hand
stripping long after the breeding season has ended. Running males can be observed as late as May for trout
(Billard, unpublished) and November for carp (Shikhshabekov, 1974). Sperm taken at the end of the reproductive
season has been shown to be of poor quality in tench
(Horoszewicz, personal communication) and in sea bass
(Billard *et al.*, 1977).

Endocrine changes during ovulation and spermiation Changes
have been observed in the hormonal status of rainbow trout
during the period of ovulation and spermiation (Breton
and Manac'H, unpublished). Estradiol 17β decreased while
t-GTH was still rising 15 days after ovulation. At the
end of spermatogenesis estradiol reaches its maximum
level and is followed by a peak in testosterone; a second
peak of estradiol occurs at the same time as t-GTH. Spermiation starts immediately afterwards.
     Data reported earlier indicate a rise of t-GTH at the
time of spermiation (Fig. 1) and oocyte maturation + ovulation (Fig. 3) in salmonids. Pituitary t-GTH is also at
its maximum in the final stages of the reproductive cycle
in salmonids (Fig. 1) and in cyprinids. A surge of plasma
c-GTH is observed at the time of ovulation in goldfish
(Breton *et al.*, 1972). In males, androgens (testosterone
and 11-ketotestosterone) increase just before spermiation
and reach a peak, which occurs earlier for testosterone
than for 11-ketotestosterone. This suggests that 11-keto-
testosterone may be more closely associated with the
phenomenon than testosterone. 11-Ketotestosterone also
rises in the winter flounder at the time of spawning
(Campbell, 1975; Campbell *et al.*, 1976).

Usually, a rise in plasma androgen and estrogen and a surge of gonadotropin occur at the time of ovulation and spermiation. Thus, hormonal requirements appear to be more important for the final stages of the cycle than for gametogenesis. This is also illustrated by the technique of hypophysation used to induce ovulation and sometimes spermiation in many species of fish (review by Pickford and Atz, 1957).

Conclusion The hormonal changes which induce ovulation are probably triggered by environmental factors which may be either abiotic (e.g. temperature) or biotic (e.g. social environment, vegetation). Thus spawning time may be linked to environmental change. However, Cushing (1969) has shown a relative fixity of spawning seasons for plaice, herring, cod and sockeye salmon. The link between spawning and environment may thus be more strict for freshwater than for marine fish which are less subject to rapid environmental change.

Conclusion

Various patterns of annual periodicity in fish reproduction can be identified in the temperate and cold zones. These patterns are not characteristic of a family. Gametogenesis can start either in autumn (roach) or in late winter (tench) in cyprinids. Most of the salmonid species exhibit gametogenesis in summer and spawn in winter, but *Thymallus* is a spring spawner. Sometimes the time for sexual maturity of male and female fish does not coincide closely; thus in the pike, ovulation in females does not always synchronise with spermiation in males. An extreme case is shown by the viviparous fish *Cymatogaster aggregata*, in which males produce sperm in the summer which is stored after copulation in the ovary of the female until the end of vitellogenesis and oocyte maturation in January. The duration of gametogenesis as well as of the breeding season is longer when gametogenesis starts in autumn. The duration of incubation should also be considered. When reproduction occurs in winter, the hatching period for some freshwater species (mainly salmonids) is delayed because of the large amount of vitellus and the low temperature. In northern Sweden, salmon spawn in the autumn and 'swim up' occurs five to six months later. The same phenomenon occurs to a lesser extent in pike which is a late winter spawner; the eggs are bigger and incubation (>200 degree-days) is longer than for spring spawning fish. If the hatching and resorption periods are considered, the reproductive cycles of winter and early spring spawners overlap and fry are released into the water when food is available.

Photoperiod and temperature are probably the main seasonal timing agents for temperate and cold water fish.

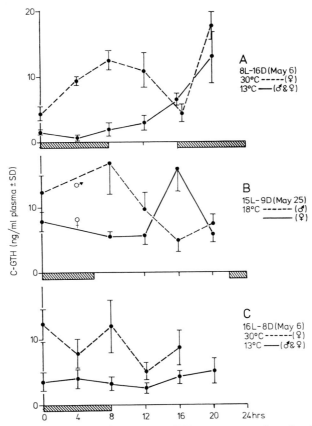

Fig. 9. Circadian rhythms in plasma GTH concentration during a 24 h period in goldfish maintained under various environmental conditions (Gillet, unpublished).

Field observations have not established which factor is essential but experimental work has given us some clues. The effects of temperature are important in cyprinids and in other groups which are often intermittent spawners. Gametogenesis may occur over a large temperature range (10 to 30°C in goldfish; Gillet, personal communication) provided that nutrition is adequate. In the natural environment, carp exhibit completion of gametogenesis at around 10-15°C and full gametogenesis may be obtained at 23°C (Gupta, 1975). Thermal requirements are more strict for ovulation in the carp, the minimum spawning temperature being 17°C (Shikhshabekov, 1974). Ovulation can be induced in goldfish at 12°C by antiestrogen implantation in the pituitary (Billard and Peter, 1977). This suggests an

action of temperature at gonadal and at central nervous
system and pituitary levels.

Photoperiod variation is probably the main environmental timing agent for some species such as salmonids and
gasterosteids (De Vlaming, 1972). The presence of a photo-
sensitive circadian rhythm, similar to that responsible
for the annual rhythm of reproduction in birds (Follett,
1973), has been demonstrated in some fish which reproduce
under long photoperiods, e.g. stickleback (Baggerman,
1972), *Heteropneustes fossilis* (Sundararaj and Vasal,
1976), and *Oryzias latipes* (Chan, 1976). The effect of
photoperiod variation on the reproductive cycle may be
mediated through an endogenous rhythm of light receptivity in some fish.

Rhythm of Reproduction in Fish Living in Tropical and Equatorial Zones

Tilapia

Field studies have been carried out on *Tilapia* in both
tropical and equatorial zones. Although environmental
factors such as photoperiod and temperature do not vary
much in the latter zone there is a seasonal reproductive
cycle. It is usually said that breeding activity corresponds to the onset of the rainy season.

Studies by Moreau (1970) in Malagasy showed that gametogenesis and breeding occurred in the austral summer and
coincided with the highest temperature; food availability
and growth rate was also maximum during this period. Observations by Kiener (1963) showed that at the same latitude,
i.e. under the same photoperiod regime, the duration of
the breeding season varied with the altitude and probably
with temperature. Data from Hyder (1970) for *T. leucostica*
in the equatorial zone are less clear. Seasonal testicular
activity was restricted to the autumn and winter. In females, GSI variation was not always correlated with the
percentage of breeding females, particularly in June or
July. Histological features showed a resting period from
July to September, as in males. This lack of gonadal activity can hardly be correlated with the known environmental
factors, although Hyder suggested a possible effect of
illumination and temperature. It should be noted, however,
that air temperature rather than water temperature was
recorded. The role of rainfall in the reproduction of
Tilapia is a matter of controversy (see Hyder, 1970, for
discussion).

Other Species of Tropical Fish

A study by Payne (1975) of a tropical stream-dwelling
cyprinid, *Barbus liberiensis* showed an annual rhythm of
reproduction with a phase of gonad recrudescence (January
to April) and a breeding season in May to July followed
by a resting period.

Rainfall was among the environmental factors important for the initiation of the various stages of reproduction; it coincided with the beginning of the breeding season. The end of the breeding season corresponded with a low somatic condition, low water temperature and declining rainfall. Gametogenesis started at the beginning of the dry period and took place during a period of slightly increasing temperature, photoperiod and ionic concentration in the water. At the end of the period of gametogenesis, no fish showed an empty foregut. A similar annual reproductive cycle was reported for *Barbustor* (Mathur, 1962), and *Heteropneustes fossilis* (Sundararaj and Vasal, 1976).

The spawning season of Caribbean reef fishes was studied by Munro *et al.*, (1973). The majority of the species studied spawn mostly in February, March and April when water temperatures are minimum at around 28°C. The Sergeant Major, *Abudefduf saxatilis*, which is eurythermal (Graham, 1972) has two main spawning periods (March and April, and July to September). Okera (1974) indicates an annual spawning season of 3-4 months duration (August to November) in the Indian ocean for *Sardinella gibbosa* and a much longer period (at least August to February) for *S. albella*. The reproductive cycles of some viviparous fish have been studied by Mendoza (1962) who found great variations in the length of the reproductive season for 3 species living in the same lake; a short reproductive cycle with one brood was noted for *Alloophorus* and *Goodea* and a breeding season extending over a period of 8 to 9 months for *Neoophorus*.

Fish Showing Continuous Reproduction Throughout the Year

In the study of Munro *et al.*, (1973), two groups of tropical fish (Carangidae and Lutjanidae) appear to spawn throughout the year. Some other species of fish can reproduce continuously provided environmental conditions are favourable; continuous reproduction was observed in the guppy *Poecilia reticulata* kept at 27°C under a constant 12L-12D photoperiod (Billard, 1966). Production of spermatozoa continued throughout the year with some variations. Young were born continuously during a 12 month period. Intervals between two parturitions were longer and more variable in winter than in summer, suggesting variations in the duration of the reproductive cycle. These variations may have been due to changes in food availability. Such a dependence of frequency of spawning on food availability has been demonstrated in *Gasterosteus aculeatus* by Wootten and Evans (1976). In other poecilid fish such as *Gambusia*, reproduction is virtually continuous under favourable conditions. Barney and Anson (1921) considered that the cyclicity of egg production was not temperature dependent but was governed by the size and metabolic potential of the mother. In another viviparous fish *Brachyrhaphis episcopi*, Turner (1938) found that breeding occurred throughout the year. Many species of marine fish appear to be continuous breeders in Indian waters (Quasim, 1973). This phenomenon of continuous reproduction in fish is well known to aquaculturists; aquarium fish kept under constant environmental conditions may exhibit continuous gametogenesis provided food availability if sufficient. Slight changes such as temperature variation or water substitution, induce spawning.

Conclusion

   In tropical or equatorial zones, most fish show an annual rhythm of reproduction. The environmental factors responsible for the various phases of the cycle are not readily identified by field observations, where many parameters vary at the same time. Gametogenesis seems to be initiated by subtle environmental changes (including slight photoperiod variations), but more dramatic events such as seasonal rainfall and flooding are required to induce ovulation as shown in the carp by Khanna (1958). This seems to have adaptive value since food availability for the young is greater during the rainy season. At the limit of the temperate and tropical zones, John (1963) noted that the reproductive cycle of *Rhinichthys osculus* in the Chiricahua mountains (Arizona) is basically regulated by photoperiod but that spawning activity was enhanced in swollen streams and flash floods.

Other Rhythms Associated with Reproduction

Spawning Rhythm

Circadian spawning activity   The daily rhythm of oviposition of the medaka (*Oryzias latipes*) (Robinson and Pugh, 1943) is well known. Eggs are laid just before dawn almost every day during the breeding season. Under artificial illumination (14L-10D), spawning occurs 30 minutes after light onset (Yamauchi and Yamamoto, 1973). Work by Egami and Nambu (1961) showed that stimuli other than photoperiodicity may also be important in determining the time of oviposition. Other species showing circadian spawning rhythms are known. The hermaphrodite, oviparous fish *Rivulus marmoratus* exhibits a 24 h rhythm of internal self-fertilization and oviposition (Harrington, 1963). During the reproductive season, *Menidia audens* lays eggs mainly in the morning (Hubbs, 1976).

   Two species of *Trichopsis* spawn at the end of the light period (Marshall, 1967) and *Pomatomus saltatrix* shows daily spawning activity near sundown (Norcross *et al.*, 1974); this is probably connected with photoperiod. Similarly, the sardine spawns in the evening (Gamulin and Hure, 1956).

Tidal and lunar spawning rhythms A daily rhythm of spawning has been reported for *Pagrus ehrenbergii* which is a tropical intermittent spawner having a 4-6 month spawning period which coincides with the rains (Stepkina, 1973). Spawning activity may occur at different times during a 24 h period and is observed when more saline and colder (denser) ocean water moves tidally to the shelf.

In Japan, the puffer *Fugu niphobles*, exhibits a lunar spawning rhythm associated with the evening tide (Nozaki *et al.*, 1976). There are other examples of lunar and tidal rhythmicity in spawning and even in gonad development (see Schwassmann, 1971 for review).

May (personal communication) has shown a rhythmic spawning activity occurring once a month in a population of *Polydactylus sexfilis*, suggesting lunar influences.

Other rhythms associated with spawning Spawning intervals of 1.9 and 2.7 days have been shown to occur in the zebrafish, *B. rerio* (Eaton and Farley, 1974).

Migration is often associated with spawning. It is unnecessary to recall the well known rhythmic reproductive migration of many species, but some fish show subtle migratory rhythms. For instance, the hogchoker *Trinectes maculatus* spawns in brackish water and the juveniles move upstream and congregate in a low-salinity nursery area where they remain during the winter. In the spring, they return to the spawning area and in September to the nursery area. This rhythmic movement continues until the fourth year; as the fish mature the migration progressively extends towards higher salinities. Thus the fish revisit the spawning area, seasonally, even when they are immature (Dovel *et al.*, 1969).

Circadian Rhythms

Circadian endocrine rhythms have been demonstrated in fish. Gonadotropin levels may vary during a 24 h period but this variation seems to depend on season, photoperiod and on the sex of the fish. In cyprinids, there is a peak of pituitary GTH concentration at 16.00 hours which disappears after pinealectomy (De Vlaming and Vodicnik, 1977). In goldfish, plasma c-GTH levels vary according to temperature, photoperiod and sex of the fish. In salmonids, a circadian rhythm in the pituitary content of GTH was shown by O'Connor (1972) but the specificity of the assay used and the reference to FSH are not convincing. Some oscillations in GTH levels in trout plasma throughout a 24 h sampling period are shown in Fig. 10; testosterone levels seem lower during the night but for both hormones any apparent differences are not statistically significant. In an experiment which involved four samplings per day, Schreck *et al.*, (1972, 1973) did not find major circadian variations for androgens and estrogens in males.

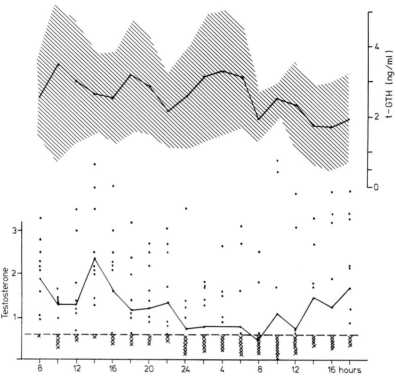

Fig. 10. Variation in T-GTH and testosterone (RIA - ng/ml) during a 24 h period in male rainbow trout (resting stage) experiment carried out in May, (Fostier *et al.*, unpublished).

Other Rhythms

A seasonal cycle was observed in the seminal vesicles and sperm ducts (Rastogi, 1969). Organs other than gonads show some variations in activity or weight during the annual cycle of reproduction. One of the more obvious changes is in the liver of the female. Usually, the liver weight rises during the resting period and gonadal recrudescence, and is minimum at the end of the spawning season. The liver is considered as the main source of lipid for vitellogenesis (Takahashi, 1974). This annual rhythmic variation of liver weight does not always appear to be connected with a specific gonadal event. In a 18 month study of brown trout, a rise of HSI was observed in December during the first year (when trout were still immature) and during ovulation in the second year. Lipid content of the body and viscera also vary with the season and usually show a peak before the onset of vitellogenesis (Rao, 1967; Shikhshabekov, 1974; Lizenko *et al.*, 1975). Usually variation is greater in males than in females (Lapin, 1973).

Seasonal variation of feeding activity (Homans and Vladykov, 1954) and circadian rhythms of food consumption
(Stepkina, 1973) have been reported. Thyroid activity is
low during the reproductive and breeding season and high
in the resting period for *Agonus cataphractus*, (Le Gall,
1969) but for a gobiid fish, *Leucopsarion petersi*, Tamura
and Honma (1973) found the highest activity in the breeding season. More accurate information has been reported
for brook trout: T_3 decreased during gametogenesis and
was minimum during the spawning period (White and Henderson, 1977).

Summary

Most teleost fish species exhibit an annual rhythm of
reproduction which is correlated with some climatic events.
In temperate zones, photoperiod and temperature variations
are the most obvious environmental factors involved in
determining the annual reproductive cycle. In tropical
zones, photoperiod and temperature are also involved but
additional factors such as the rainy season contribute.
Annual rhythms of reproduction are also observed in fish
at the equator and the timing agents are then less obvious
(temperature, food availability, rainfall, sunshine, etc.
....).
    The reproductive cycle consists of two parts, gameto-
genesis and spawning; these include ovulation, oviposition,
spermiation + sperm release and spawning behaviour.
    Gametogenesis appears to depend on regular and long
term changes (circannual variations of temperature or
photoperiod for some species: both temperature and photoperiod, sometimes with interaction, for others). The presence of a circadian, light-sensitive phase has been demonstrated in fish suggesting that induction of gametogenesis
by photoperiod change may be mediated through an endogenous daily rhythm. Endocrine studies have shown that relatively low and regularly increasing amounts of GTH are
required for gametogenesis indicating a secretion type
'tonic release' compatible with regular change.
    Spawning seems to require more specific stimuli, even
for fish with continuous reproduction, and appears to be
the most critical period of the reproductive cycle. These
stimuli are sometimes a thermal shock in temperate zones
(cyprinids, pike), and flooding, rainfall and probably
many others in tropical and equatorial zones. Endocrine
studies indicate that ovulation usually occurs after a
GTH surge associated with other hormonal changes, and
with variations in gonadal receptivity to gonadotropin.
These are probably brought about by climatic stimuli. It
is thus likely that spawning is triggered by climatic
events. This may be an adaptation to the environment
since hatching occurs soon after reproduction (at least
for small egg breeders) and since vulnerable larvae

require adequate environmental conditions. However, this
'trigger' concept may be associated with endogenous rhythms; such rhythms exist for ovulation-oviposition in
O. latipes, for instance. Mammalian species have evolved
an endogenous cyclical mechanism for ovulation, perhaps
because homeothermy and gestation and suckling processes
render the young less sensitive to the immediate environment. Nevertheless, fish should not be regarded as being
more primitive in their reproductive strategies than mammals, on the contrary, they have developed modes of reproduction which are adaptive, varied and often original.

Acknowledgement

Thanks are due to our colleagues, Mrs. Escaffre and
Reinaud, Drs. Fostier, Gillet, Jalabert and Weil, who
provided us with personal results and criticised the
manuscript. Mrs. Daifulu improved the English presentation
of the manuscript. Graphs were drawn or redrawn by Miss
Briand. Laboratory work was partly supported by grants
from the E.E.C., E.D.F. and C.N.R.S.

References

Ahsan, S.N. (1966). *Can. J. Zool.* **44**, 149-159.
Baggerman, B. (1972). *Gen. Comp. Endocrinol. Suppl.* **3**, 466-476.
Bara, G. (1960). *Rev. Fac. Sci. Univ. Istanbul, Ser B.* **25**, 49-91.
Barnabe, G. (1976). Thesis, Univ. Sci. Tech. du Languedoc, 426pp.
Barney, R.L. and Anson, B.J. (1921). *Anat. Rec.* **22**, 317-335.
Beach, A.W. (1959). *Can. J. Zool.* **37**, 615-625.
Billard, R. (1966). Thèse 3-ème cycle Fac. Sci. Lyon, 83pp.
Billard, R. (1968). *C.R. Acad. Sci.* **266**, 2287-2290.
Billard, R. (1969). *Ann. Biol. Anim. Bioch. Biophys.* **9**, 251-271.
Billard, R. and Peter, R.P. (1977). *Gen. Comp. Endocrinol.* **32** (2), 213-220.
Billard, R., Breton, B. and Du Bois, M.P. (1971). *C.R. Acad. Sci.* **272**, 981-983.
Billard, R., Solari, A. and Escaffre, A.M. (1974). *Ann. Biol. Anim. Bioch. Biophys.* **14**, 87-104.
Billard, R., Dupont, J. and Barnabe, G. (1977). *Aquaculture,* **11**, (4) 363-367.
Breton, and Billard, R. (1977). *Ann. Bio. anim. Biophys, Biochim.* **17** (3A) 331-340.
Breton, B., Billard, R., Jalabert, B. and Kann, G. (1972). *Gen. Comp. Endocrinol.* **18**, 463-468.
Brusle, J. and Brusle, S. (1975). *Rev. Trav. Inst. Pêches. Marit.* **39**, 313-320.
Campbell, C.M., Walsh, J.M. and Idler, D.R. (1976). *Gen. Comp. Endocrinol.* **29**, 14-20.
Campbell, C.M. (1975). PhD Thesis Memorial Univ. of Newfoundland, St John's Newfoundland, Canada.
Caporiccio, B. (1976). Thèse 3-ème cycle. Univ. Sci. Tech. du. Languedoc.
Chan, K.K.S. (1976). *Can. J. Zool.* **54**, 852-856.

O'Connor, J.M. (1972). *Comp. Biochem. Physiol.* **43A**, 739-746.
Contini, A. and Donato, A. (1973). *Mem. Biol. Mar. Oceanogr.*, *Messina*, **3**, (6), 173-184.
Crim, L.W., Watts, E.G. and Evans, D.M. (1975). *Gen. Comp. Endocrinol.* **27**, 62-70.
Cushing, D.H. (1969). *J. Cons. Int. Explor. Mer.* **33**, 81-92.
Dovel, W.L., Mihursky, J.A. and McErlean, A.J. (1969). *Chesapeake Sci.* **10**, 104-119.
Eaton, R.G. and Farley, R.D. (1974). *Copeia*, **1974**, 195-204.
Egami, N. and Nambu, M. (1961). *J. Fac. Sci. Univ. Tokyo. Ser. IV*, **9**, 263-278.
Egami, N. and Hyodo-Taguchi, Y. (1967). *Exp. Cell. Res.* **47**, 665-667.
Eleftheriou, B.E., Norman, R.L. and Summerfelt, R. (1968). *Steroids*, **11**, 89-95.
Escaffre, A.M. and Billard, R. (1976). *Cah. Lab. Hydrobiol. Montreau*. **3**, 43-46.
Follett, B.K. (1973). *J. Reprod. Fert.* **Supp 19**, 5-18.
Fournier, P. and Magnin, E. (1975). *Nat. Can.* **102**, 181-187.
Le Gall, S. (1969). *Vie et Milieu* **1A**, 153-234.
Gamulin, T. and Hure, J. (1956). *Nature* **177**, 193-194.
Graham, J.B. (1972). *Physiol. Zool.* **45**, 1-13.
Gupta, S. (1975). *J. Fish. Biol.* **7**, 775-782.
Harrington, R.W. (1963). *Physiol. Zool.* **36**, 325-341.
Heins, D.C. and Clemmer, G.H. (1976). *J. Fish. Biol.* **8**, 365-379.
Henderson, N.E. (1962). *Can. J. Zool.* **40**, 631-641.
Homans, R.E.S. and Vladykov, V.D. (1954). *J. Fish. Res. Bd Canada* **11**, 535-542.
Honma, Y. and Tamura, E. (1962). *Jap. J. Ichthyol.* **9**, 135-152.
Hubbs, C. (1976). *Copeia* **1976**, 386-388.
Hyder, M. (1970). *J. Zool. Lon.* **162**, 179-195.
Jalabert, B., Breton, B. and Billard, R. (1974). *Ann. Biol. anim. Bioch. Biophys.* **14**, 217-228.
John, K.R. (1963). *Copeia* **1963**, 286-291.
Jones, J.W. (1940). *Proc. Roy. Soc. Ser B. Biol. Sci.* **128** (853) 499-509.
Khanna, D.V. (1958). *Indian J. Fisheries.* **5**, 282-290.
Kiener, A. (1963). *Poissons, Peche et pisciculture a Madagascar*. C.I.F.T. Ed. 160pp.
Labios, M.D. (1965). *Gen. Comp. Endocrin.* **5**, 207-221.
Lahaye, J. (1972). *Rev. Trav. Inst. Pêches Marit.* **36**, 191-207.
Lapin, V.I. (1973). *J. Ichthyol.* **13**, 262-274.
Lizenko, Y.I., Siderov, V.S. and Potapova, O.I. (1975). *J. Ichthyol.* **15**, 465-472.
Mackay, N.J. (1973). *Aust. J. Zool.* **21**, 67-74.
Marshall, J.A. (1967). *Animal Behaviour*, **15**, 510-513.
Mathur, D.S. (1962). *Zoologica Poloniae*, **12**, 131-144.
Merriner, J.V. (1975). *Fish. Bull.* **74**, 18-26.
Mieskowska, A. and Jasinski, A. (1973). *Bull. Acad. Pol. Sci.* **21**, 395-398.
Moreau, J. (1970). Thesis 3-eme cycle Fac. Sci. Paris. 112pp.
Moser, G.H. (1967). *J. Morph.* **123**, 329-353.
Munro, J.L., Gaut, V.C., Thompson, R. and Reeson, P.H. (1973). *J. Fish*

Biol. **5**, 69-84.

Norcross, J.J., Richardson, S.L., Massman, W.H. and Joseph, E.B. (1974). *Trans. Amer. Fish. Soc.* **103**, 477-497.

Nozaki, M. *et al.* (1976). *Zool. Mag. Tokyo,* **85**, 156-158.

Okera, W. (1974). *J. Fish. Biol.* **6**, 801-812.

Ozon, R. (1972). *In: Steroids in nonmammalian vertebrates,* (D.R. Idler, ed.), 329-389 and 390-414. Academic Press, London.

Payne, A.I. (1975). *J. Zool.* **176**, 247-269.

Pickford, G. and Atz, J.W. (1957). *The physiology of the pituitary gland of fishes.* Academic Press.

Pitt, T.K. (1966). *J. Fish. Res. Bd. Can.* **23**, 651-672.

Quasim, S.Z. (1973). *Indian J. Fish.* **20**, 166-181.

Rao, K.S. (1967). *J. Mar. Biol. Ass. India,* **9**, 303-322.

Rastogi, R.K. (1969). *Acta anat.* **72**, 624-639.

Robinson, E.J. and Rugh, R. (1943). *Biol. Bull.* **84**, 115-125.

Sandor, T. and Idler, D.R. (1972). *In: Steroids in Nonmammalian Vertebrates,* (D.R. Idler, ed.), 6-36, Academic Press, London.

Sangalang, G.B. and Freeman, H.C. (1974). *Biol. Reprod.* **11**, 429-435.

Schmidt, P.J. and Idler, D.R. (1962). *Gen. Comp. Endocrinol.* **2**, 204-214.

Schreck, C.B., Lackey, R.F. and Hopwood, M.L. (1972). *Copeia,* **1972**, 865-868.

Schreck, C.B., Lackey, R.F. and Hopwood, M.L. (1973). *J. Fish. Biol.* **5**, 227-230.

Schreck, C.B. and Hopwood, M.L. (1974). *Trans. Am. Fish. Soc.* **103**, 375-378.

Schwassmann, H.O. (1971). *In: Fish Physiology,* (Hoar and Randall, eds.), VI, 371-428, Academic Press, London.

Shikhshabekov, M.M. (1974). *J. Ichthyol.* **14**, 79-87.

Siciliano, M.J. (1972). *J. Fish. Biol.* **4**, 131-140.

Smith, C.L. and Young, P.H. (1966). *Calif. Fish. and Game.* **52**, 283-292.

Stepkina, M.V. (1973). *J. Ichthyol.* **13**, 641-649.

Summerfelt, R.C. and Minckley, C.O. (1969). *Trans. Amer. Fish. Soc.* **98**, (3) 444-453.

Sundararaj, B.I. and Vasal, S. (1976). *J. Fish. Res. Bd Canada,* **33**, 959-973.

Takahashi, S. (1974). *Bull. Jap. Soc. Sci. Fish.* **40**, 847-857.

Tamura, E. and Honma, Y. (1973). *Bull. Jap. Soc. Sci. Fish.* **39**, 1003-1011.

Turner, C.C. (1938). *Biol. Bull.* **75**, 56-65.

Turner, P.R. and Summerfelt, R.C. (1971). *Amer. Fish. Soc. Spec. Pub.* **No. 8**, 107-119.

De Vlaming, V.L. (1972). *J. Fish. Biol.* **4**, 131-140.

De Vlaming, V.L. (1974). *In: Control of Sex in Fishes,* (C.B. Schreck, ed.), 13-83. Extension Division Virginia Polytechnic Institute and State University, Blacksburg. Virginia 24061.

De Vlaming, V.L. and Vodicnik, M.J. (1977). *J. Fish. Biol.* **10**, 73-86.

White, B.A. and Henderson, N.E. (1977). *Can. J. Zool.* **55**, 475-481.

Webe, J.P. (1968). *Can. J. Zool.* **46**, 1221-1234.

Wingfield, J.C. and Grimm, A.S. (1977). *Gen. Comp. Endocrinol.* **31**, 1-11.

Wootton, R.J. and Evans, G.W. (1976). *J. Fish. Biol.* **8**, 385-395.
Yamauchi, K. and Yamamoto, K. (1973). *Ann. Zool. Japan* **46**, 144-153.

AN INTERPRETATION OF SOME ENDOCRINE RHYTHMS IN FISH

T.H. SIMPSON

Marine Laboratory, Aberdeen

Introduction

Despite the rapid growth in our understanding of endocrine processes and the remarkable advances in endocrine methodology, the goal of a direct, unequivocal correlation of endocrine changes with particular events in the life cycle of fish remains elusive. This elusiveness reflects the fact that the links in the extended chain between the initiating 'instruction' to an endocrine gland and the effect of the hormone on a target tissue or process are themselves subject to variable modification by environmental changes. Thus, for example, in the chain, neural instruction → hypothalamus → release/inhibitor factor → pituitary → thyrotrophin → thyroid gland → thyroid hormone → target, the responsiveness of each of the tissues is variably affected by the concentrations of other hormones, and expectedly by temperature, while two of the transmitters, thyrotrophin and thyroid hormone are subject to a temperature and nutrition dependent metabolic breakdown. In view of the complexities of systems of this kind, it is not surprising that there can be no authoritative description of fish activity rhythms in terms of causal endocrine events; it may, nevertheless, be useful to attempt to draw parallels. The reproductive endocrinology of fish is reviewed in this Symposium by Billard and Breton (1978) and pituitary-pineal relationships by Matty (1978). The purpose of the present communication is to review the roles of thyroid and adrenocortical hormones and to assess the significance of rhythmic changes in their production to the life cycle of fish.

Thyroid Hormones

Thyroid function in fish has been the subject of recent reviews by Gorbman (1969) and by Y-A. Fontaine (1975); for reviews of the earlier literature, see Pickford and Atz

(1957), Dodd and Matty (1964) and Barr (1965). The primary thyroid hormone, thyroxine (T_4) is synthesised within the thyroid follicles, which may be encapsulated, as in selachians, or relatively diffuse, as in most teleosts, by mechanisms which are similar to those which operate in higher vertebrates (Taurog, 1974; Greer and Haiback, 1974). Among the teleosts, control of T_4 production is exercised by thyrotrophin (TSH); such control appears to be tenuous in selachians (Lewis and Dodd, 1976) and has not been demonstrated in cyclostomes (Larsen and Rosenkilde, 1971; Matty *et al.*, 1976). Hypothalamic regulation of TSH production in teleosts has been established (see Ball *et al.*, 1972; Peter, 1973 for reviews); in most, though perhaps not all species, the control is exerted by an inhibitor factor (Bromage *et al.*, 1976). Among the cyclostomes and elasmobranchs, no such neuroendocrine control of TSH seems anatomically possible (Gorbman, 1969).

The role of thyroid hormones in fish is still far from clear, indeed the existence of marked seasonal cycles of thyroid gland activity perhaps provides the strongest presumptive evidence of the importance of thyroid hormone. The bulk of the extremely conflicting evidence (Gorbman, (1969) is that T_4 does not increase oxygen consumption in fish; Hochachka (1962) has suggested that the conflicting results on oxygen uptake can be reconciled if, as his results suggest, thyroid hormones act particularly to activate the pentose phosphate cycle of carbohydrate metabolism (see also Brown, 1960; Hochachka, P.W. and Hayes, F.R., 1962). Recent work by Ruhland (1969, 1971) on oxygen consumption in *Aequides latifrons* has convincingly demonstrated an increase in consumption after treatment with T_4 and a decrease after thiorea. Moreover, T_4 has been shown to uncouple oxidation and phosphorylation in muscle in the brown trout at much lower concentrations than in mammals (Massey and Smith, 1968) and to induce swelling in mitochondria isolated from dogfish (Greif and Alfano, 1964).

The roles of thyroid hormones in protein and lipid metabolism in fish and on the processes involved in linear growth are more clear. At moderate concentrations, thyroid hormones promote the incorporation of leucine into goldfish and trout muscle (Thornburn and Matty, 1963) and of glycine into newly synthesised integumentary guanine in rainbow trout (Matty and Sheltawy, 1967). More recently, T_4 has been found to reduce nitrogen excretion in fed rainbow trout (Smith and Thorpe, 1977). Higher levels are associated with a negative nitrogen balance (Hoar, 1958; Woodhead, 1975) and with a thinning of the dermis and epidermis (La Roche and Leblond, 1952). T_4 increases bone plate formation in thyroxinised sturgeon (Gerbilsky and Saks, 1947), skeletal growth and calcification in rainbow trout (La Roche *et al.*, 1966) and the incorporation of

radio-labelled sulphur into the skeleton of trout (Barrington and Rawdon, 1967). Thyroid hormones have been shown
to reduce abdominal fat levels in the rainbow and in brook
trout (Baraduc, 1954; Barrington et al., 1961; Narayansingh
and Eales, 1975), to lower serum lipid levels (Takashima
et al., 1972) and to raise muscle free fatty acids (Narayansingh and Eales, 1975). A general picture emerges of
thyroid hormones promoting growth (Higgs et al., 1976) at
low concentrations and, at higher concentrations, of mobil-
ising lipids and proteins for use as energy sources.

The relationship between hormonal status and swimming
behaviour in fish has been reviewed by Woodhead (1975).
T_4 has been shown to stimulate locomotor activity in salmonid species (Hoar et al., 1952, 1955), in the guppy
(Sage, 1968), stickleback (Baggerman, 1962) and in the
cod Gadus morhua (Woodhead, 1970). That the converse of
this relationship may also hold is suggested by the studies of Fontaine and Leloup (1959, 1962) and of Higgs and
Eales (1971). However, although the exercised trout had
more active thyroids than the sedentary fish, it seems
likely that the heightened activity of the gland was a
response to an increase in general metabolism leading to
a temporary reduction in plasma hormone levels and reduced negative feedback at the hypothalamic level. Godin
et al., (1974), have recently reported the extremely interesting, paradoxical observation that swimming activity,
upstream orientation and aggressive behaviour in yearling
Atlantic salmon were significantly reduced after treatment
with T_4 or tri-iodothyronine (T_3) such reductions would
be expected to facilitate the downstream migration of
smolts.

Thyroid hormones have been shown to affect central
nervous function in fish, increasing their responsiveness
to external stimulation (Hoar et al., 1955), sensitising
optically evoked potentials in the mid brain and modifying
the potentials in the olfactory bulb evoked by chemical
stimulation of the olfactory apparatus (Hara et al., 1965,
1966; Oshima and Gorbman, 1966a, 1966b; Hara and Gorbman,
1967).

Adrenocortical Hormones

The functional morphology of steroidogenic tissues in
lower vertebrates has been reviewed by Lofts and Bern
(1972). There appears to be no single piscine homologue
of the adrenal cortex of mammals. Among the teleosts, the
presumed homologue, the interrenal tissue is distributed
in the haemopoietic tissues of the head kidney and closely
associated with the postcardinal veins (Nandi, 1962); insufficient data are available on the functional anatomy
of the interrenal tissues of non-teleostean bony fish to
permit any general description for these orders. In chon-
drichthyean fishes, the interrenal is organised into a

discrete, encapsulated gland, located posteriorly; in the
Agnatha, the homologue of the adrenal cortex has not been
defined and the source of the low concentrations of corticosteroids in their blood plasma has not yet been established (Seiler *et al.*, 1970; Weisbart and Idler, 1970).
   The occurrence and biosynthesis of corticosteroids in
fish has been critically reviewed by Idler and Truscott
(1972) and the biological effects of these compounds by
Chester Jones *et al.*, (1972) and by Fontaine (1975). The
importance of corticosteroids to the osmoregulatory problems encountered by migratory fish has been discussed by
Woodhead (1975). The present brief survey will be confined
to the so-called 'glucocorticoids' and their effect on
fish metabolism. In almost all orders of fish, the principal corticosteroids identified in blood plasma or in
the incubation products of interrenal tissue with appropriate radioactive precursors, are the 17-hydroxylated
C_{21} steroids, cortisol and cortisone; corticosteroids reported in small quantities for some species include 11-
deoxycortisol and corticosterone. The elasmobranch fish
differ in having no interrenal 17-hydroxylase, the principal corticosteroids of these species being 1α-hydroxy-
corticosterone and perhaps corticosterone. The relative
abundance of these steroids in blood plasma differs between species and appears also to vary in individual specimens with successive stages of the life cycle. As in
higher vertebrates, secretion of corticosteroids has been
shown to be controlled by corticotrophin (ACTH) in teleosts and selachians (see Chester Jones *et al.*, 1973 for
review) and in chondrostean fish (Barannikova, 1974).
The observation that extracts of the hypothalamus of goldfish caused a release of ACTH from the pituitary *in vitro*
(Sage and Purrot, 1969) suggests that in this species at
least, a hypothalamic release factor (CRF) exerts ultimate
control of corticosteroidogenesis.
   In mammals, glucocorticoids promote gluconeogenesis,
the synthesis of glucose from non-carbohydrate precursors,
e.g. glycerol and amino acids, inhibit the utilisation of
glucose by tissues (Landau, 1965) and depress skeletal
growth (see Daughaday *et al.*, 1975 for review). Studies
of the effects of hypophysectomy and of the direct administration of corticosteroids together with field observations of changes in plasma steroid levels occurring at
different stages of the life cycle suggest that corticosteroids exert similar effects in fish; these effects
are best documented in the teleosts. Hypophysectomy resulted in a depletion of liver glycogen in the eels *Anguilla anguilla* (Hatey, 1951a, 1951b) and *A. rostrata*
(Butler, 1968) and in hypoglycaemia in the hypophysectomised bullhead, *Ictalurus melas* (Chidambaram *et al.*, 1973);
these effects were reversed by the administration of corticosteroids or ACTH. Field observations are in general

support of the conclusions of these manipulative studies.
Thus for example, extensive catabolism of body reserves
is associated with high plasma corticosteroid titres and
an increase in liver glycogen levels in the upstream migration of *Oncorhynchus nerka* (Idler and Clemens, 1959;
Idler *et al.*, 1959a, 1959b; Chang and Idler, 1960).

In teleosts, the main substrate used to support the
increased gluconeogenesis induced by corticosteroids
appears to be muscle protein, although lymphoid tissue
and leucocytes may contribute (Pickford *et al.*, 1970).
Storer (1967) found that exogenous cortisol induced a loss
of parietal muscle in the goldfish and Smith and Thorpe
(1977) reported that cortisol increased nitrogen excretion
in the starved, though not in the fed rainbow trout. The
general catabolism necessary to support gluconeogenesis
is commonly reflected in a decrease in somatic growth.
Storer (1967) reported a reduction in growth in goldfish
after treatment with cortisol and Pickford *et al.*, (1970)
found that a similar treatment caused a reduction in the
body weight of *Fundulus heteroclitus*. Ball (quoted in
Chester Jones *et al.*, 1969) and Ball and Ensor (1969)
noted that cortisol reduced both the weight and length
of hypophysectomised *Poecilia latipinna* and impaired the
natural regeneration of amputated dorsal fins. Changes in
skeletal growth may be expected to reflect changes in the
secretion of somatotrophin. Olivereau and Olivereau (1968)
reported that interrenalectomy in the eel was followed by
an increase in the activity of pituitary somatotrophs.
These underwent regression after treatment with cortisol.
Analogous results have been observed in the stickleback,
Gasterosteus aculeatus by Leatherland and Lam (1971). They
reported that repeated injections of ACTH caused significant reductions in the body length and in the size of
pituitary somatotrophs of winter-phase fish. The fact
that cortisol was found to be less effective than ACTH
is perhaps not surprising since the injection regime adopted for the former compound would not be expected to maintain elevated plasma corticoid levels for the duration of
the experiment.

In some teleost species, corticosteroids (or ACTH)
appear to induce glycogenolysis, a conversion of glycogen
to glucose, in addition to gluconeogenesis. Swallow and
Fleming (1966, 1969) found that administration of ACTH
to fasted, intact *Tilapia mossambica* failed to restore
liver glycogen to the fed levels but increased its rate
of turnover.

The role of corticosteroids in glucose metabolism in
other Classes of fish is much less clear. In the hagfish,
Myxine glutinosa, hypophysectomy was not attended by any
significant effects (Falkmer and Matty, 1966). Administration of cortisol or ACTH to fasting lampreys caused
initial increases in liver glycogen and plasma glucose

levels (Bentley and Follett, 1965); prolonged treatment
caused subsequent decreases in liver glycogen. In chond-
richthyean fish there is only slight evidence of a gluco-
neogenetic role for corticosteroids. Idler *et al.*, (1969)
(c.f. Hartmann *et al.*, 1944) reported that interrenalectomy of the skate, *Raja erinacea* had no effect on glycogen
levels. Patent (1970) observed that corticosterone induced
hyperglycaemia in *Squalus acanthias* and in *Hydrolagus
colliei* and increased liver glycogen levels in the former.
In view of the much more striking increase in liver lipoly-
sis, it was suggested that, in chondrichthyeans, corticosteroids may be more concerned with lipid than with
carbohydrate metabolism.

Endocrine Rhythms

Unlike terrestial vertebrates, most species of fish
experience, annually, periods of starvation caused by
either a 'voluntary' reduction of food intake or by a
seasonal reduction in the availability of food. Thus, for
example, salmonid fish do not feed during the lengthy
freshwater phase of their journey to the spawning beds
and in both the herring (Wood, 1958; Parrish and Saville,
1965) and haddock (Homans and Vladikov, 1954), there is
a marked reduction in food intake during the later stages
of gonad maturation. In general, somatic growth and development of the gonads may be regarded as being alternative
processes (Swift, 1955; Iles, 1964) which must be expected
to have endocrine correlates. The changes in thyroid gland
and interrenal activity reported as occurring during spawning migrations (reviewed by Woodhead, 1975) are largely
deduced from histological observations and must, in any
case, be regarded as composites of changes due to environmental factors, stress and pituitary-gonad effects, with
changes due to inanition. Osborn and Simpson (1974) and
Simpson *et al.*, (unpublished observations) found that in
the rainbow trout, food deprivation which was reflected
in a reduction in the mean body lipid content from 8% to
2%, resulted in a significant drop in serum T_4 concentrations, an even more marked reduction in serum T_3 and
a concurrent, significant rise in serum cortisol. T_3 in
the rainbow trout originates largely as a hepatic metab-
olite of T_4. The observation that food deprivation was
associated with a decrease in T_3/T_4 ratios and that the
starved fish were notably less active and less responsive
than the controls, suggests that the lowered T_4 levels
served to reduce metabolism and to moderate corticoid-
induced catabolism. Such a response would clearly be of
adaptive value during periods of food scarcity. Leatherland *et al.*, (1977) have also observed reductions of plasma T_4 levels in starved rainbow trout and have reported
that the fish responded with different T_4 concentrations
to diets of different fat contents.

Although short term effects of stress have been studied (see Chavin, 1973 for review) less is known of long term reactions. Osborn and Simpson (1972, 1974) found that trawl or transport stress resulted in marked reductions in plasma T_4 and T_3 concentrations and in T_3/T_4 ratios in the plaice, *Pleuronectes platessa* and in the rainbow trout; depressed levels persisted for periods of up to 30 days. Since exogenous TSH restored normal thyroid hormone levels, the reduction of T_4 and T_3 was considered to be a response to an inhibition of TSH by the elevated plasma corticosteroid levels; corticosteroid-induced reductions in thyroid follicular cell heights have been reported in *Astyanax mexicanus* (Rasquin and Atz, 1952) and in *Carassius* (Chavin, 1956). Reductions in plasma T_4 and T_3 levels and, implicitly, in T_3/T_4 ratios have been noted also in the elasmobranch *Scyliorhinus canicula* during captivity (Lewis and Dodd, 1976). Longlasting elevations of plasma cortisol concentrations have been reported in stressed rainbow trout (Simpson, 1976); as in the case of the endocrine changes induced by food deprivation, stress-induced changes are of clear survival value.

It is perhaps inevitable that studies of endocrine rhythms in fish should have been related to the reproductive cycle rather than to more subtle environmental changes. More unfortunate is the fact that the only methods available to the earlier investigators provided data which is now regarded as being of doubtful value, if not actually misleading. Histological observations provide only an indication of the activity of a gland and do not distinguish between the different biochemical conditions which result when endocrine glands of constant activity are secreting into organisms in which the hormone is subject to different rates of hepatic degradation or peripheral utilisation. The endocrine status of the organism is best defined in terms of the concentrations of hormones in the fluids bathing the different target sites; the early analytical methods for determining these concentrations were often of doubtful specificity (see Sandor and Idler, 1972 for a critique of steroid methodology).

The earlier literature on thyroid rhythms in fish, reviewed by Matty (1960), Swift (1960) and Dodd and Matty (1964) suggests that although thyroid hormones may have no obligatory role in piscine reproduction, there is nevertheless a clear parallel between reproductive and thyroid rhythms, thyroid activity being at its maximum at or just before spawning. Changes in thyroid activity have been variously interpreted as responses to annual changes in temperature and day length (Swift, 1960), to the direct effect of changing gonadal hormones on the gland (Matty, 1960) and to the increased metabolic demand imposed by heightened locomotor activity during gonad maturation (Hoar, 1957). More recent work has failed to elucidate

this problem. Bromage and Sage (1968) reported that the
cycle of thyroid activity associated with gestation cycles
in *Poecilia reticulata* occurred under conditions of constant temperature and day length. Direct effects of steroids on the thyroid gland were shown in experiments by
Singh (1969) in which 'estrogen', testosterone and cortisone were found to restore thyroidal uptake of radioiodine
in hypophysectomised catfish, *Mystus vittatus*. In papers
presented in this symposium, Osborn *et al.*, (1977) and
Osborn and Simpson (1977) report an annual rhythm in plasma T_4 and T_3 determined by gas-liquid chromatography, in
rainbow trout and in the plaice; at no sampling time was
there any difference between the hormone levels of males
and females or between mature and immature fish. It thus
seems likely that the rhythm in plasma hormone levels is
one which permits rather than directs the major biochemical processes involved in gonad maturation. White and
Henderson (1977) reported that plasma T_3 levels in *Salvelinus fontinalis* increase during gonad development and
decline sharply before spawning. Similar results are described in reports by the author and his colleagues cited
above. It is of interest that spawning is associated with
a surge in plasma gonadotrophin (Billard and Breton, this
symposium) and that T_4 was found to reduce the release of
gonadotrophin in *Poecilia reticulata, in vitro* and *in vivo*
(Sage and Bromage, 1970). It is possible therefore that
the decline in plasma T_4 and T_3 before spawning is one
which permits the necessary surge in gonadotrophin.

Seasonal changes in the activity of interrenal tissue
and in the levels of plasma corticosteroids have been reported for many species of fish (see Woodhead, 1975 for
review). These changes are perhaps best documented in the
salmonid fish and have been associated with the anadromous
spawning migration (Hane and Robertson, 1959; Robertson
and Wexler, 1959; Schmidt and Idler, 1962). Similar changes
occur, during the period of gonad development, in non-
migratory salmonids including rainbow trout (Robertson
et al., 1961), in *Salvelinus leucomaensis* (Honma and Tamura, 1965), *Coregonus lavaretus* (Fuller *et al.*, 1974) and
in the winter flounder, *Pseudopleuronectes americanus*
(Campbell *et al.*, 1976). Leach and Taylor (1977) report
a circannual rhythm in serum glucose concentrations in
Fundulus heteroclitus, maximum activities coinciding with
gonad development. The large scatter in serum cortisol
determinations precluded any correlation of cortisol with
hyperglycaemia. The possibility that activation of the
interrenal is specifically associated with gonadogenesis
is contra-indicated by the fact that immature fish have
been reported to show such an activation at the same time
as mature members of the same population. Woodhead and
Woodhead (1965) observed interrenal hyperplasia in immature
as in mature *Gadus morhua,* in the Barents Sea in early

winter. Recently, Wingfield and Grimm (1977) reported
that similar maximum values in the annual cycle of plasma
cortisol, measured by radioimmunoassay, were attained at
peak spawning time in immature and mature plaice, *Pleuronectes platessa*. The annual rhythm of adrenocortical hormone status, like the rhythm in thyroidal status, may
therefore be regarded as one which permits the biochemical
processes involved in the development of the gonad but is
not specifically directed by gonadogenesis.

It is of interest to compare the changes in thyroid and
adrenocortical status occurring during stress and food deprivation with those occurring during gonad development.
In the former, increases in corticosteroids are accompanied by lowered thyroid hormone levels; it was suggested
earlier that a reduction of metabolic activity and a mod-
eration of corticosteroid-induced catabolism would be the
expected result. In contrast, during gonad development increases in thyroid hormone levels occur concurrently with
increases in plasma corticosteroids. This association
would be expected to increase metabolism and corticosteroid controlled catabolism; such processes are implicitly
involved in the depletion of body reserves and their trans-
location to the gonad during maturation.

Considerable interest, stimulated by the work of Meier
(see Meier, 1975 for review), has recently been attached
to possible circadian rhythms in the endocrine status of
fish. Boehlke *et al.*, (1966), using methods which are open
to some criticism, found a diurnal rhythm in plasma corticosteroids in the channel catfish, *Ictalurus punctatus;*
maximum levels were reported as occurring in the afternoon.
Chavin (1973) reported a bimodal, diurnal rhythm in plasma
cortisol in the fasted goldfish, maxima occurring at 1200
and 1600 hours; regrettably, the analytical procedure used
for the steroid determinations cannot be regarded as being
highly specific. Preliminary results from this laboratory
(Simpson *et al.*, unpublished) indicate a statistically
significant diurnal variation in fed rainbow trout. Using
a proven radioimmunoassay technique for the determination
of cortisol we have shown a single maximum in cortisol
levels occurring during the night, as in higher vertebrates
(see Halberg, 1969 for review). During September, a single
rhythm was observed for male and female fish. During March
however only the male fish showed a statistically significant rhythm with a maximum at 0400 hours, as in September;
cortisol levels in the females were lower than in the males
and appeared to be arhythmic. Lee and Meier (1967) and de
Vlaming and Sage (1972) have shown that injection of prolactin into various species of cyprinodont fish results
in fat deposition or fat mobilisation, depending on the
time of injection relative to the onset of light. Natural
rhythms of prolactin concentrations in the goldfish have
been reported by Leatherland and McKeown (1973) and by

Spieler and Meier (1975), and in juvenile *Oncorhynchus
nerka* by Leatherland *et al.*, (1974). Moreover, Meier *et
al.*, (1971) and Joseph and Meier (quoted in Meier, 1972)
have shown that in two *Fundulus* species, maintained under
constant illumination, daily injections of cortisol served
to entrain the response to prolactin, the length of the
time interval between the cortisol and prolactin injection
determining the nature and magnitude of the response, fat
deposition or fat mobilisation.
T_4 also has been found to phase fattening responses to
prolactin in *F. chrysotus* under conditions of constant
illumination; a single injection of T_4 served to entrain
the response to prolactin for many subsequent cycles
(Meier, 1970). Any seasonal variation in fattening of
fish, caused by the prolactin mechanism, would require
that both hormones show a diurnal rhythm and that at least
one should show a seasonal change in its timing. Diurnal
variations in plasma, T_4 levels have been reported in bird
(Sadovsky and Bensadoun, 1971; Newcomer, 1974). The half-
life of T_4 in rainbow trout, like that in birds is much
lower than in mammals and short enough for variations in
the secretory rate of the thyroid gland to find expression
in diurnally varying plasma levels. Osborn *et al.*, (1977,
this symposium) report diurnal variations in T_4 levels in
fed rainbow trout, maxima occurring 1 hour after sunset
in September and 5 hours after sunset in March. These
observations add weight to the suggestion that in fish,
as in other vertebrates physiological changes may be controlled by the relative, temporal phasing of hormones as
well as by their absolute levels.

Acknowledgements

    The author wishes to acknowledge his debt to his pres-
ent and former colleagues, in particular, Drs. R.S. Wright
R.H. Osborn, A.F. Youngson, R. Johnstone and Miss S.V. Hun
for invaluable discussions and for making available unpublished data.

References

Baggerman, B. (1962). *Gen. Comp. Endocrinol.*, **1**, 188-205.
Ball, J.N. and Ensor, D.M. (1969). *In: La Specificite Zoologique des
    Hormones Hypophysaires et de leurs Activite*. Colloq. Intern.
    Centre Natl. Rech. Sci., Paris, **177**, 215-224.
Ball, J.N., Baker, B.I., Olivereau, M. and Peter, R.E. (1972). *Gen.
    Comp. Endocrinol.*, supp. **3**, 11-21.
Baraduc, M.M. (1954). *C.r. hebd. Seanc. Acad. Sci., Paris*, **283**, 728-
    730.
Barannikova, I.A. (1974). *Gen. Comp. Endocrinol.*, **22**, 367.
Barr, W.A. (1965). *Oceanogr. Mar. Biol. Ann. Rev.*, **3**, 257-298.
Barrington, E.J.W. and Rawdon, B.B. (1967). *Gen. Comp. Endocrinol.*,
    9, 116-128.
Barrington, E.J.W., Barron, N. and Piggins, D.J. (1961). *Gen. Comp.*

Endocrinol., **1**, 170-178.
Bentley, P.J. and Follett, B.K. (1965). *J. Endocr.*, **31**, 127-137.
Billard, R. and Breton, B. (1978). This volume, pp. 31-54
Boehlke, K.W., Church, R.L., Tiemeier, O.W. and Eleftheriou, B.E. (1966). *Gen. Comp. Endocrinol.*, **7**, 18-21.
Bromage, N.R. and Sage, M. (1968). *J. Endocr.*, **41**, 303-311.
Bromage, N.R., Whitehead, C. and Brown, T.J. (1976). *Gen. Comp. Endocrinol.*, **29**, 246.
Brown, W.D. (1960). *J. Cell. Comp. Physiol.*, **55**, 81-85.
Butler, D.G. (1968). *Gen. Comp. Endocrinol.*, **10**, 85-91.
Campbell, C.M., Walsh, J.M. and Idler, D.F. (1976). *Gen. Comp. Endocrinol.* **29**, 14-20.
Chang, V.M. and Idler, D.R. (1960). *Can. J. Biochem. Physiol.*, **38**, 553-558.
Chavin, W. (1956). *J. expl. Zool.*, **153**, 259-279.
Chavin, W. (1973). *In: Responses of Fish to Environmental Changes.* (W. Chavin, ed.), Charles C. Thomas, Springfield, Illinois, 199-238.
Chester Jones, I., Bellamy, D., Chan, D.K.O., Follett, B.K., Henderson, I.W., Phillips, J.G. and Snart, R.S. (1972). *In: Steroids in Non-mammalian Vertebrates* (D.R. Idler, ed.), Academic Press, London and New York, 415-468.
Chester Jones, I., Chan, D.K.O., Henderson, I.W. and Ball, J.N. (1969). *In: Fish Physiology, Vol. III, The Endocrine System,* (W.S. Hoar and D.J. Randall, eds.), Academic Press, London and New York, 321-376.
Chester Jones, I., Ball, J.N., Henderson, I.W., Sandor, T. and Baker, B.I. (1973). *In: Chemical Zoology* (M. Florkin, ed.), Academic Press, London and New York, VIII, pp. 844-921.
Chidambaram, S., Meyer, R.K. and Hasler, A.D. (1973). *J. expl. zool.*, **184**, 75-80.
Daughaday, W.H., Herington, A.C. and Phillips, L.S. (1975). *Ann. Rev. Physiol.*, **37**, 211-244.
Dodd, J.M. and Matty, A.J. (1964). *In: The Thyroid Gland,* (R. Pitt-Rivers and W.R. Trotter, eds.), Butterworth, London and Washington, D.C., 303-356.
Falkmer, S. and Matty, A.J. (1966). *Gen. Comp. Endocrinol.*, **6**, 334-346.
Fontaine, M. and Leloup, J. (1959). *C.r. hebd. Seanc. Acad. Sci., Paris,* **249**, 343-347.
Fontaine, M. and Leloup, J. (1962). *Gen. Comp. Endocrinol.*, **2**, 317-322.
Fontaine, Y-A. (1975). *In: Biochemical and Biophysical Perspectives in Marine Biology* (D.C. Malins and J.R. Sargent, eds.), Academic Press, London and New York, 139-197.
Fuller, J.D., Scott, D.B.C. and Fraser, R. (1974). *J. Endocr.*, **63**, 24p.
Gerbilsky, N.L. and Saks, M.G. (1947). *Dok Akad. Nauk SSSR*, **55**, 663-666.
Godin, J.G., Dill, P.A. and Drury, D.E. (1974). *J. Fish Res. Bd Can.* **31**, 1787-1790.
Gorbman, A. (1969). *In: Fish Physiology, Vol. II, The Endocrine System,* (W.S. Hoar and D.J. Randall, eds.), Academic Press, London

and New York, 241-265.
Greif, R.L. and Alfano, J.A. (1964). *Gen. Comp. Endocrinol.*, **4**, 339-342.
Greer, M.A. and Haiback, H. (1974). *In: Handbook of Physiology*. Vol. VII (M.A. Greer and D.H. Solomon, eds.), Amer. Physiol. Soc., 135-146.
Halberg, F. (1969). *Ann. Rev. Physiol.*, **31**, 675-725.
Hane, S. and Robertson, O.H. (1959). *Proc. Nat. Acad. Sci. USA*, **45**, 886-893.
Hara, T.J. and Gorbman, A. (1967). *Comp. Biochem. Physiol.* **21**, 185-200.
Hara, T.J., Ueda, K. and Gorbman, A. (1965). *Gen. Comp. Endocrinol.* **5**, 313-319.
Hara, T.J., Gorbman, A. and Ueda, K. (1966). *Proc. Soc. Exptl. Biol. Med.* **122**, 471-475.
Hartman, F.A., Lewis, L.A., Brownwell, K.A., Angerer, C.A. and Sheldon, F.F. (1944). *Physiol. Zool.*, **17**, 228-238.
Hatey, J. (1951a). *C.r. Seanc. Soc. Biol.*, **145**, 172-175.
Hatey, J. (1951b). *C.r. Seanc. Soc. Biol.*, **145**, 315-318.
Higgs, D.A. and Eales, J.G. (1971). *Can. J. Zool.*, **49**, 1255-1269.
Higgs, D.A., Donaldson, E.M., Dye, H.M. and McBride, J.R. (1976). *J. Fish. Res. Bd Can.*, **33**, 1585-1603.
Hoar, W.S. (1957). *In: The Physiology of Fishes*. (M.E. Brown, ed.), Academic Press, London and New York, Vol. 1., 245-286.
Hoar, W.S. (1958). *Can. J. Zool.*, **34**, 113-121.
Hoar, W.S., MacKinnon, D. and Redlich, A. (1952). *Can. J. Zool.*, **30**, 273-286.
Hoar, W.S., Keenleyside, M.H.A. and Goodall, R.G. (1955). *Can. J. Zool.* **33**, 428-439.
Hochachka, P.W. (1962). *Gen. Comp. Endocrinol.*, **2**, 499-505.
Hochachka, P.W. and Hayes, F.R. (1962). *Can. J. Zool.*, **40**, 261-270.
Homans, R.E.S. and Vladykov, V.D. (1954). *J. Fish. Res. Bd Can.*, **11**, 535-542.
Honma, Y. and Tamura, E. (1965). *Japan. Soc. Sci. Fish.*, **31**, 867-882.
Idler, D.R. and Clemens, W.A. (1959). *In: International Pacific Salmon Fisheries Commission Progress Report*, New Westminster, B.C., Canada, 80pp.
Idler, D.R. and Truscott, B. (1972). *In: Steroids in Nonmammalian Vertebrates* (D.R. Idler, ed.), Academic Press, London and New York, 127-212.
Idler, D.R., Ronald, A.P. and Schmidt, P.J. (1959a). *J. Am. Chem. Soc.*, **81**, 1260.
Idler, D.R., Ronald, A.P. and Schmidt, P.J. (1959b). *Can. J. Biochem. Physiol.* **37**, 1227-1238.
Idler, D.R., O'Halloran, M.J. and Horne, D.A. (1969). *Gen. Comp. Endocrinol.*, **13**, 303-306.
Iles, T.D. (1964). *J. Cons. perm. int. Explor. Mer*, **29**, 166-188.
Landau, B. (1965). *Vitamins Horm.*, **23**, 2-60.
La Roche, G. and Lablond, C.P. (1952). *Endocrinology*, **51**, 524-530.
La Roche, G. Woodall, A.N., Johnson, C.L. and Halves, J.E. (1966). *Gen. Comp. Endocrinol.*, **6**, 249-266.
Larsen, L.O. and Rosenkilde, P. (1971). *Gen. Comp. Endocrinol.*, **17**,

94-104.
Leach, G.J. and Taylor, M.H. (1977). *Comp. Biochem. Physiol.*, **55A**, 217-223.
Leatherland, J.F. and Lam, T.J. (1971). *J. Endocr.*, **51**, 425-436.
Leatherland, J.F. and McKeown, B.A. (1973). *J. interdiscipl. Cycle Res.* 4, 137-143.
Leatherland, J.F., McKeown, B.A. and John, T.M. (1974). *Comp. Biochem. Physiol.*, **47A**, 821-828.
Leatherland, J.F., Cho, C.Y. and Slinger, S.J. (1977). *J. Fish Res. Bd Can.*, **34**, 677-682.
Lee, R.W. and Meier, A.H. (1967). *J. expl. Zool.*, 166, 307-316.
Lewis, M. and Dodd, J.M. (1976). *Gen. Comp. Endocrinol.*, **29**, 258.
Lofts, B. and Bern, H.A. (1972). *In: Steroids in Nonmammalian Vertebrates*, (D.R. Idler, ed.), Academic Press, London and New York, 37-117.
Massey, B.D. and Smith, C.L. (1968). *Comp. Biochem. Physiol.*, **25**, 241-255.
Matty, A.J. (1960). *Symp. Zool. Soc. Lond.*, **2**, 1-15.
Matty, A.J. (1978). This volume, pp.21-30
Matty, A.J. and Sheltawy, M.J. (1967). *Gen. comp. Endocrinol.*, **9**, 473 (abstr.).
Matty, A.J., Tsuneki, K., Dickhoff, W.W. and Gorbman, A. (1976). *Gen. Comp. Endocrinol.*, **30**, 500-516.
Meier, A.H. (1970). *Proc. Soc. Exp. Biol. Med.*, **133**, 1113-1116.
Meier, A.H. (1972). *Gen. Comp. Endocrinol. Supplement* **3**, 449-508.
Meier, A.H. (1975). *In: Hormonal Correlates of Behaviour* (B.E. Eleftheriou and R.L. Sprott, eds.), Plenum Press, New York, 469-549.
Meier, A.H., Trobec, T.N., Joseph, M.M. and John, T.M. (1971). *Proc. Soc. Exp. Biol. Med.*, **137**, 408-415.
Narayansingh, T. and Eales, J.G. (1975). *Gen. Comp. Endocrin.*, **52**, 407-412.
Nandi, J. (1962). *Univ. Calif. (Berkeley) Publ. Zool.*, **65**, 129-212.
Newcomer, W.S. (1974). *Gen. Comp. Endocrinol.*, **24**, 65-73.
Olivereau, M. and Olivereau, J. (1968). *Z. Zellforsch. mikrosk. Anat.*, **84**, 44-58.
Osborn, R.H. and Simpson, T.H. (1972). *Gen. Comp. Endocr.* **13**, 613.
Osborn, R.H. and Simpson, T.H. (1974). *Cons. Perm. Int. Explor. Mer* CM 1974 Gear and Behaviour Comm. **B9**.
Osborn, R.H. and Simpson, T.H. (1977). *J. Fish Biol.* In Press.
Osborn, R.H., Simpson, T.H. and Youngson, A.F. (1977). *J. Fish. Biol.* In press.
Oshima, K. and Gorbman, A. (1966a). *Gen. Comp. Endocrinol.* **7**, 398-409.
Oshima, K. and Gorbman, A. (1966b). *Gen. Comp. Endocrinol.* **7**, 482-491.
Parrish, B.B. and Saville, A. (1965). *Oceanogr. Mar. Biol. Ann. Rev.*, **3**, 323-373.
Patent, G.J. (1970). *Gen. Comp. Endocrinol.*, **14**, 215-242.
Peter, R.E. (1973). *Amer. Zool.*, **13**, 743-755.
Pickford, G.E. and Atz, J.W. (1957). *The Physiology of the Pituitary Gland of Fishes*. N.Y. Zool. Soc.
Pickford, G.E., Pang, P.K.T., Weinstein, E., Torretti, J., Hendler, E. and Epstein, F.H. (1970). *Gen. Comp. Endocrinol.*, **14**, 524-534.
Rasquin, R. and Atz, E.H. (1952). *Zoologica, N.Y.* **37**, 77-87.

Robertson, O.H. and Wexler, B.C. (1959). *Endocrinology*, **65**, 225-238.
Robertson, O.H., Krupp, M.A., Thomas, S.F., Favour, C.B., Hane, S. and Wexler, B.C. (1961). *Gen. Comp. Endocrinol.*, **1**, 473-484.
Ruhland, M.L. (1969). *Experentia*, **25**, 944-945.
Ruhland, M.L. (1971). *Can. J. Zool.*, **49**, 423-425.
Sadovsky, R. and Bensadoun, A. (1971). *Gen. Comp. Endocrinol.*, **17**, 268-274.
Sage, M. (1968). *Gen. Comp. Endocrinol.* **10**, 304-309.
Sage, M. and Bromage, N.R. (1970). *Gen. Comp. Endocrinol.*, **14**, 137-140.
Sage, M. and Purrot, R.J. (1969). *Zeit, M. and Purrot, R.J.* (1969). *Zeit. fur Vergl. Physiol.* **63**, 85-90.
Sandor, T. and Idler, D.R. (1972). *In: Steroids in Nonmammalian Vertebrates*, (D.R. Idler, ed.), Academic Press, London and New York, 6-36.
Schmidt, P.J. and Idler, D.R. (1962). *Gen. Comp. Endocrinol.*, **2**, 204-214.
Seiler, K., Seiler, R. and Sterba, G. (1970). *Acta. Biol. Med. Ger.*, **24**, 553.
Simpson, T.H. (1976). *Proc. R. Soc. Edinb.*, (B), **75**, 241-252.
Singh, T.P. (1969). *Gen. Comp. Endocrinol.*, **12**, 556-560.
Smith, M.A.K. and Thorpe, A. (1977). *Gen. Comp. Endocrinol.*, **32**, 400-406.
Spieler, R.E. and Meier, A.H. (1975). *J. Fish. Res. Bd Can.*, **33**, 183-186.
Storer, J.H. (1967). *Comp. Biochem. Physiol.*, **20**, 939-948.
Swallow, R.L. and Fleming, W.R. (1966). *Am. Zool.*, **6**, 562.
Swallow, R.L. and Fleming, W.R. (1969). *Comp. Biochem. Physiol.*, **28**, 95-106.
Swift, D.R. (1955). *J. exp. Biol.*, **32**, 751-764.
Swift, D.R. (1960). *Symp. Zool.*, *Soc. Lond.*, **2**, 17-27.
Takashima, F., Hibaya, T., Phan-van Ngan and Aida, K. (1972). *Bull. Jap. Soc. Sci. Fish.*, **38**, 43-49.
Taurog, A. (1974). *In: Handbook of Physiology* (M.A. Greer and D.H. Solomon, eds.), Amer. Physiol. Soc. 101-134.
Thornburn, C.C. and Matty, A.J. (1963). *Comp. Biochem. Physiol.*, **8**, 1-12.
de Vlaming, V.L. and Sage, H. (1972). *Contr. in Marine Sci.*, **16**, 59-63.
Weisbart, M. and Idler, D.R. (1970). *J. Endocr.*, **46**, 29.
White, B.A. and Henderson, N.E. (1977). *Can. J. Zool.*, **55**, 475-481.
Wingfield, J.C. and Grimm, A.S. (1977). *Gen. Comp. Endocrinol.*, **31**, 1-11.
Wood, J.D. (1958). *Can. J. Biochem. Physiol.*, **36**, 833-838.
Woodhead, A.D. (1975). *In: Oceanography and Marine Biology*. (H. Barnes, ed.), Allan and Unwin. London, 287-382. Vol. 13.
Woodhead, A.D. and Woodhead, P.M.J. (1965). *Int. Comm. N.W. Atl. Fish. Spec. Publ.* No. **6**, 691-715.
Woodhead, P.M.J. (1970). *J. Fish. Res. Bd Can.*, **27**, 2337-2338.

NOCTURNALISM VERSUS DIURNALISM - DUALISM WITHIN FISH INDIVIDUALS

LARS-OVE ERIKSSON*†

Messaure Biological Station,
S-96036 Messaure, Sweden
and
Max-Planck-Institut für Verhaltensphysiologie,
D-8131 Erling-Andechs, FR Germany.

Introduction

Since the days of the Cold Spring Harbor symposium in 1960 the number of studies of circadian rhythms has increased rapidly. This has also been the case in fish. However, apart from the fact that most fish show a diel rhythm in all parameters measured and that light has been found to be the main environmental variable affecting the rhythmic patterns, the results presented so far are by no means as clear as those obtained from similar studies in other organisms. The reports do not even always agree in such an apparently evident thing as the phasing of the activity patterns; to look in the literature of laboratory assays to find out whether a particular fish species is nocturnal or diurnal might be hazardous indeed.

For example, Swift (1962, 1964) reported that the brown trout (*Salmo trutta* L.) was a diurnal species, while Chaston (1968, 1969) said that it was strictly nocturnal.

Since Swift investigated lake brown trout and Chaston brown trout from a creek it would be possible that there is some conspicuous difference between populations from streams and lakes. However, Müller (1969) has shown that individuals of stream-dwelling trout at the Arctic circle are nocturnal *or* diurnal, depending on time of year. To make the confusion complete, according to my observations (Eriksson, 1973) brown trout are basically crepuscular,

*Footnote: Supported by the Swedish Natural Research Council, the Alexander von Humboldt-Stiftung and the Max-Planck-Institut für Verhaltensphysiologie at Erling-Andechs.

†Present address: Department of Ecological Zoology, University of Umeå, S-901 87 Umeå, Sweden.

TABLE 1

Reports indicating a dual phasing of the diel activity rhythms in fish

Species	diurnal	Reported as crepuscular	nocturnal	Difference intra-	Difference interindividual	Reference
Salmo salar	x					Hoar, 1942
	(x)	x	(x)	(x)		Eriksson, 1973, 1975
	x	x	x		x	Varanelli and McCleave, 1974
						Richardson and McCleave, 1974
Salmo trutta	x					Swift, 1962, 1964
			x			Chaston, 1968, 1969
	x		x	x		Müller, 1969
	(x)	x	(x)	(x)		Eriksson, 1973, 1975
Oncorhynchus nerka	x		x			Byrne, 1971
Phoxinus phoxinus	x		x		x	Jones, 1956
	x					Müller, this volume
Lota lota	x		x		x	Penaz, 1975
	(x)		x	x		Wikgren, 1955
	x		x	x		Müller, 1970a, 1973

Lota lota	x	x	Kroneld, 1976
Cottus poecilopus		x	Andreasson, 1969
	x	x	Müller, 1970b, this volume
Cottus gobio	x	x	Andreasson, 1973 Müller, this volume
Myoxocephalus quadricornis	x	x	Westin, 1971

twilight-active.

A number of reports claiming or indicating the existence of such a dualistic phasing of the diel activity rhythms in fish is listed in Table 1. The reports include laboratory assays only, and fall into three categories:

(i) Opposite patterns found in two or more independent investigations within one species (*Salmo salar*, *S. trutta*, *Phoxinus phoxinus*, and *Cottus poecilopus*)

(ii) Opposite patterns *between fish individuals*, found in one and the same investigation (*Salmo salar*, *Phoxinus phoxinus*)

(iii) Opposite patterns *within individual fish*, depending on the time of the experiment (*Salmo salar*, *S. trutta*, *Oncorhynchus nerka*, *Phoxinus phoxinus*, *Lota lota*, *Myoxocephalus quadricornis*, *Cottus poecilopus* and *C. gobio*)

Certainly, there are several problems related to the convenient methods of fish activity measurements. It has been shown that differences in recording techniques may well be responsible for several contradictions regarding the phasing of locomotor activity rhythms in fish (Eriksson, 1975).

In order to review critically evidence for dualistic (diurnal and nocturnal) phasing in fish, I exclude data that are related to (*a*) measurements using a single method, (*b*) measurements of activity of more than one individual at one time, and (*c*) observations of changes during the ontogeny of the individual*. This leaves the conclusion that a capacity for dualistic phasing of the diel activity rhythm in fish has been established in the following species so far: *Salmo salar*, *S. trutta*, *Lota lota*, *Cottus poecilopus* and *C. gobio*.

The inversion phenomenon (Eriksson, 1973) on an *individual* basis has been observed mainly during long-term investigations at high northern latitudes. Generally, it is strictly bound to time of year, with diurnalism during the height of the winter. Fish showing a seasonal inversion of the diel activity pattern can be divided into two categories:

(i) Crepuscular fish (*Salmo salar* and *S. trutta*). In natural light conditions, activity maxima in these fish are always closely related to dawn and dusk, independent of season. Apparent diurnalism or nocturnalism will depend on (*a*) time of year (distance between dawn and dusk), and (*b*) the amount of

---

*Footnote: A dualistic capacity is a more or less instantaneous switching from diurnal to nocturnal activity and vice versa. Ontogenetic changes like those shown for sockeye salmon by Byrne (1971) have therefore been disregarded.

activity appearing at either side of the dawn and
dusk maxima.
(ii) Biphasic fish (*Lota, lota, Cottus peocilopus*
and *C. gobio*). Traditionally considered as strictly
nocturnal species, these fish often become diurnal
during midwinter when kept in subarctic light conditions. The change from nocturnalism to diurnalism
and vice versa is rather drastic and takes place via
a period of apparent arhythmicity (Müller, 1970a,b,
and this volume; Andreasson, 1973; Kroneld, 1976)..

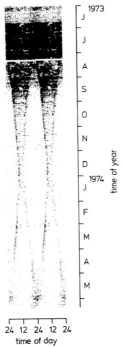

Fig. 1. 373 days record of locomotor activity of a young sea trout,
kept in natural light at the arctic circle (Messaure, Swedish Lapp-
land, 66° 42' N). Passage of the trout through an infra-red light-
beam causes a mark on event-recorder paper. Each day's record has
been glued below the other and the record has been duplicated for
better clarity. Note the more pronounced diurnal activity in autumn
(August-September) compared with spring (April-May).

The apparently somewhat different studies considered
here have a common problematical background. Firstly it
was found that sea-trout parr kept in natural light conditions at the Arctic circle throughout the year, tended
to be much more diurnal in early autumn than under corresponding photoperiods in spring (Fig. 1). Secondly, the diel

rhythms of trout subjected to artificial light conditions
were found to vary considerably. The variation seemed to
depend on time of year.

Brown bullheads (*Ictalurus nebulosus* LeSueur) were
introduced to our laboratory because of their general con-
venience as laboratory animals and their very clear activity pattern (Fig. 2). However, subjected to low artificial
light intensities the otherwise prominent nocturnal activity disappeared and the fish tended to become diurnal.

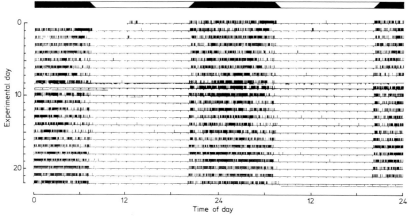

Fig. 2. Typical diel activity pattern of a brown bullhead. Performance
as in Fig. 1.

Dualism Within Brown Trout Individuals

Experimental Procedure

One year old brown trout parr between 9 and 11 cm total
length were obtained from the salmon hatchery at Heden,
Norrbotten County in northernmost Sweden, in May, 1973.

They were transferred to Messaure Biological Station,
situated close to the Arctic circle (66° 42' N). Six fish
were placed in individual circular channels, 14 cm wide
by 15 cm deep, outer diameter 64 cm. The tanks were provided with sand, gravel and some larger stones, the latter
to force the trout to choose position at a distance of
10-15 cm from the single photocell window of the tank.
A group of reserve fish were held together in a separate
tank.

Experimental and reserve fish were offered a continuous
flow (5-7 l/min) of filtered creek water and thus exposed
to the natural seasonal variations in water temperature
(range 3.5-17.7°C during the period May-September, range
0.8-3.5°C during the period October-April). All fish were
fed 3-5 times weekly, with dry salmon food and living
worms.

The experiments were carried out during the period

May, 1973 to June 1974. Photoperiod treatments of the
experimental animals are shown in Fig. 3. Fish were subjected to natural light conditions throughout the year,
except for four periods of 1 month each: June, October,
December 20- January 20, and April.

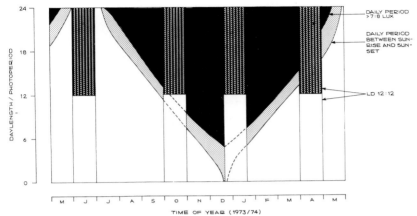

Fig. 3. Schematic description of the experiments to test the annual
differences in phasing of the diel locomotor activity in brown trout.
Actual photoperiod offered the fish at different times of the year
to be read on the left.

During these four periods the single window of the small
experimental room was covered by black plastic and an
artificial light-dark regime LD 12:12 (200:0.1 lx) was
substituted for natural light conditions. Locomotor activity was measured by means of infra-red light (λ>850nm)
photocell systems. The number of interruptions of the
beam by fish were recorded by a print-out counter, print-
ing each hour (Eriksson (1973, 1975) gives a detailed
description of the method). Two experimen-
tal fish died in March 1974. They were substituted by reserve fish. As these had been subjected to the same experimental procedures as experimental fish and did not react
differently from experimental fish, data produced by them
have been used in the statistical treatments of April
values.

The 'Annual Response Curve' of Brown Trout (Fig. 4)

In relation to the natural situation an artificial
light-dark cycle of LD 12:12 in June means a shortening
of the photoperiod by 12 h. Under these circumstances,
brown trout change from an apparent 'nocturnal' phasing
of the diel locomotor activity rhythm prior to the treatment to a diurnal phasing in artificial light. In autumn,
the treatment means practically no change of photoperiod

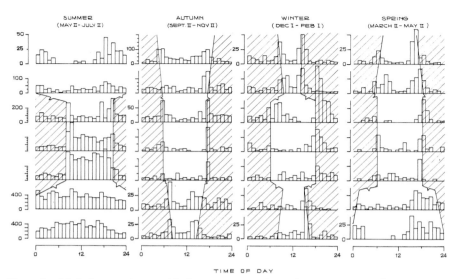

Fig. 4. Diel locomotor activity pattern in a brown trout after transference from natural light to LD 12:12. The staples are 10-day means of hourly number of passages. Transfer took place after the second 10-day period at each season. Note the differences in numbers of passages at different seasons, depending on differences in water temperature. The dark period of the light-dark cycle is marked by the hatched area.

in relation to previous natural circumstances. The crepuscular activity of trout in natural light at this time of year is replaced by light-on and light-off peaks in artificial light. In contrast to the case in summer, the same fish subjected to LD 12:12 in winter become more or less pronouncedly nocturnal. LD 12:12 in winter is equal to about 7 h prolongation of the photoperiod in relation to the natural circumstances. The April test came somewhat late, and the treatment meant a slight shortening of the photoperiod. The fish shown in Fig. 4, and several others, showed a less consistent reaction than during other times of the year. In this case almost complete nocturnalism during the first ten days of the test period was followed by progressively developing diurnalism. The results from all 6 test animals have been summarized in Fig. 5. Young sea-trout have 72, 50, 33 and 55 percent of their diel activity during the 12 h of artificial day in summer, autumn, winter and spring, respectively. The differences are statistically significant between summer and autumn-winter, and between autumn and winter respectively (U-test, two-tailed, after Wilcoxon, Mann and Whitney (Snedecor and Cochran, 1967)).

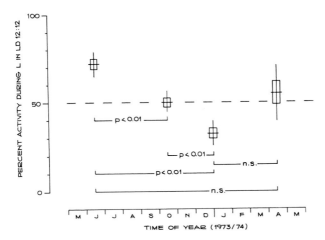

Fig. 5. Summary of the resulting gross activity distribution of 6 experimental fish in LD 12:12 at different seasons.

Atlantic salmon have diel and seasonal patterns of activity very similar to those of brown trout, at least at the Arctic circle (Eriksson, 1973, 1975). Richardson and McCleave (1974) studied locomotor activity of salmon parr in LD 12:12. In all, they tested 88 fish for a period of about 14 days each. 87 fish were rhythmic, of which 35 were diurnal, 20 nocturnal, and 32 crepuscular. I have re-arranged these data on phasing with respect to time of year of the test (Table 2, see also Richardson and McCleave, 1974, table 1).

TABLE 2

Phasing of diel locomotor activity rhythms in Atlantic salmon parr after transference from nLD to LD 12:12 (data from Richardson and McCleave, 1974)

Test season	diurnal	nocturnal	crepuscular
Summer (June-August)	30	1	11
Autumn	-	-	-
Winter (Feb.-March)	4	12	7
Spring (April-May)	1	7	14

There are very clear differences in the phasing between animals tested in winter and in summer. Almost all animals in the summer were diurnal, and most animals tested in winter were nocturnal. The spring data also agree with those from brown trout in that most animals were crepus-

cular, although we perhaps could expect more clearly diurnal animals in April-May.

The large seasonal differences in the phasing of the
diel locomotor activity rhythm in brown trout and Atlantic
salmon are certainly astonishing. However, regular seasonal changes of circadian parameters are well-known (Aschoff
and Wever, 1962; Aschoff, 1969; Daan and Aschoff, 1975 for
reviews). Birds transferred to constant light conditions
have systematic seasonal changes at least regarding activity time (Pohl, 1972; Gwinner, 1975a). General effects of
prior experience of the experimental animal on its free-
running circadian rhythm have been discussed by Pittendrigh
and Daan (1976). They show convincingly that, for example,
exposure to long or short photoperiods prior to exposure
to DD will give different frequency of the circadian rhythm
in several species of nocturnal rodents (so called 'after-
effects'). The difference in phasing of brown trout in
LD 12:12 at different times of the year may in part be
regarded as 'after-effects'. At least they mirror pronounced seasonal differences in the physiological state
of the fish. If it is assumed that under natural light
conditions brown trout will react differently in phasing
of activity between shortening and lengthening photoperiods, then the spring-to-autumn differences of phasing in
brown trout at the Arctic circle, as shown in Fig. 1, are
readily explained.

Dualism Within Brown Bullhead Individuals

Experimental Procedure

Bullheads 5.0-12.4 g in weight (c. 1 yr old) were obtained from a hatchery close to Hamburg and transferred
to Erling-Andechs. They were placed into separate 15 1
rectangular transparent plastic tanks, 17 cm wide by 18 cm
deep by 45 cm long. Each tank was supplied with filtering-
aeration equipment, a sand bottom and a shelter to hide
in. The water temperature was kept constant at 20± 1°C.
Food (living *Tubifex* spp.) was supplied to the tanks *ad
lib.*, and vitamins were added to the water weekly. During
the experiments, brown bullheads grew rapidly and animals
in the longest experiments gained 300-450% in weight in
one year. Locomotor activity of the fish was measured by
means of an infra-red light-photocell equipment ($\lambda > 750$ nm)
with the beam crossing the tank at about 20 cm from the
front and about 5-7 cm above the bottom. In a few cases,
feeding activity was recorded electromechanically by
means of a feed counter, monitoring the attempts of the
fish to take *Tubifex* penetrating the net walls of a hol-
der (Eriksson *et al.*, in prep.).

The events of fish crossing the light beam or taking
food were collected by a process computer, restored by
computer and stored on tape on a one-minute basis for
later analysis (consult Daan, 1976 for a thorough descrip-

tion of the storage-analysis system).

In the main experiment described here, fish were kept in a light-dark regime (fluorescent light tubes) with light for 11 h 03 min, with dark for 10 h 33 min, and with dawn and dusk for 1 h 33 min, respectively. All fish were offered a light-to-dark ratio of about 1000:1, but with different light intensities according to the following:

    (i) L = 120 lx   D = 0.1 lx   6 individuals
   (ii)   10   0.01   3   "
  (iii)   1   0.001   6   "

The Relativity of Nocturnalism in Brown Bullheads

During the first period of exposure to light of different intensities, there were clear, but small differences between the experimental groups (Fig. 6). The trend was

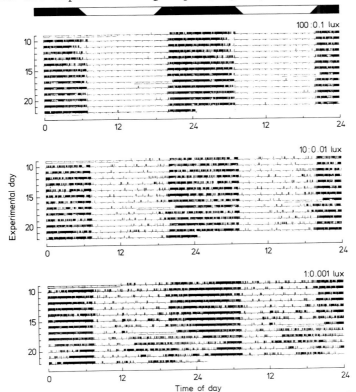

Fig. 6. The direct effect of light intensity on the diel activity distribution in brown bullheads. Note changes in appearance of activity during the bright period. Performance as in Fig. 1.

towards higher activity during the bright period with decreasing light intensity. During the first month of the

test the percent nocturnal activity was 95, 86, and 68 in
group (i), (ii) and (iii) respectively.
    With time, most animals in the lowest light intensities
became diurnal. The change from nocturnalism to diurnalism
was rather abrupt in two cases, as exemplified in Fig. 7.

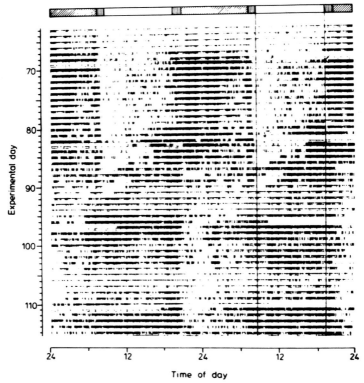

Fig. 7. Phase-shift of locomotor activity rhythm in a brown bullhead
after prolonged exposure to a low light intensity (1:0.001 lx). Performance as in Fig. 1.

One part of the nocturnal activity band started progressively earlier in the afternoon and after a transient
process it locked in a diurnal phase.
    In 3 cases, the change from nocturnalism to diurnalism
took place more slowly (Fig. 8), and in the following
sequence:
        (i) Phototactic behaviour changed: frequent obser-
        vations showed that these fish gave up use of their
        shelters long before any significant change in the
        diel activity patterns became obvious. During the
        day, they rested on the bottom for most of the time.
        (ii) A diurnal pattern was built up with the noc-

turnal pattern remaining relatively undisturbed.
(iii) Nocturnal activity diminished gradually.

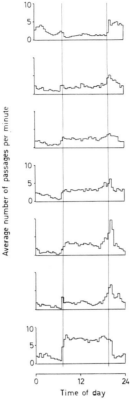

Fig. 8. Gradual change of the diel activity distribution of a brown bullhead after prolonged exposure to a low light intensity. The staples show half-monthly mean values of average numbers of passages per minute. The start of the twilight periods have been marked by lines.

Fig. 9 summarizes the result of up to 13.5 mo of exposure to the three different light intensities. I have defined as nocturnal, patterns where more than 67% of total activity occurs during the dark period. Diurnal patterns are those with less than 33% activity during the dark, and patterns with 33-67% activity during the dark period I have called indifferent. Night has been defined as the time between 8 a.m. and 8 p.m. This is certainly an oversimplification, because the start of dusk was at 7.30 a.m. and start of dawn at 7.15 p.m. As will be shown later, this will lead to an underestimation of the amount of diurnal activity.
    With these restrictions in mind, all fish subjected to

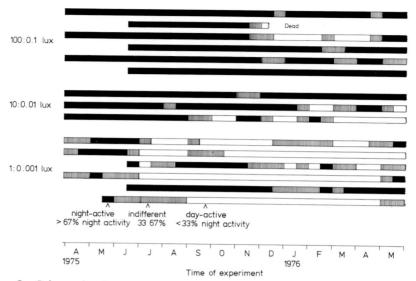

Fig. 9. Schematic description of the effect of long-term exposure of brown bullheads to three different light conditions on their diel activity distribution.

the highest light intensity showed a very stable nocturnal pattern during the first 6-9 mo of the experiment. After this time, 4 out of 5 fish which survived to the end of the experiment sometimes showed shorter or longer periods of indifferent patterns. One fish was diurnal for about 3 mo.

In the lowest light intensity, 4 out of 6 animals became diurnal within a few months of the experiment. One fish was rather indifferent throughout and one remained nocturnal during most of the period.

The animals in the medium light intensity tended to be somewhat less stably nocturnal than animals in the highest intensity of illumination. On the other hand they were clearly more stable than low-intensity animals, at least during the first half-year.

Expressed as the percent of the experimental days that individual animals were nocturnal, indifferent or diurnal, respectively, low intensity animals were significantly less often nocturnal, more often indifferent, and diurnal for longer periods than animals at high light intensities (Fig. 10). Although the number of experimental animals in the middle light intensity was low, the nonlinearity of the relationship between light intensity and phasing is very clearly indicated. A critical threshold of light intensity where brown bullheads will become diurnal within a limited period of time seems to lie somewhere close to the regime 1:0.001 lx.

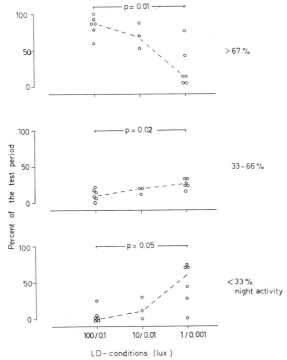

Fig. 10. Summary of the long-term effects of light intensity upon the diel distribution of locomotor activity in brown bullheads. The figure shows the percent of time ($6\frac{1}{2}$ - 14 mo) that individual animals were nocturnal, indifferent, and diurnal, respectively.

Relations Between the Light-dark Cycle and the Activity Rhythms in Nocturnal and in Diurnal Bullheads

There are large differences in the relation between activity patterns and the light-dark cycle between nocturnal and diurnal brown bullheads (Fig. 11).

Nocturnal animals have an activity time clearly limited by the dark period. Diurnal bullheads, on the other hand, have an activity period ranging from before dawn, over the whole bright period and into darkness again. In addition, a small nocturnal pattern often appears in the middle of the night.

The onset and end of activity are not always as pronounced in diurnal animals as they are in nocturnal. It has been possible to collect data from 185 activity onsets and 214 activity ends. Between 75 and 80 percent of all these clearly distinguishable activity onsets and activity

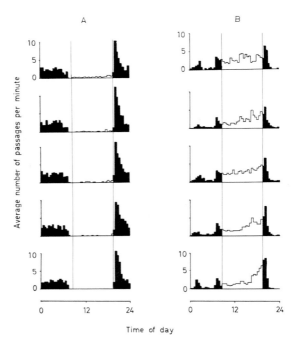

Fig. 11. The relation between the light-dark cycle and the diel activity distribution in nocturnal (A) and a diurnal (B) bullhead. Activity during the bright period (excluding dawn and dusk) is shown by white bars; otherwise performance as in Fig. 8.

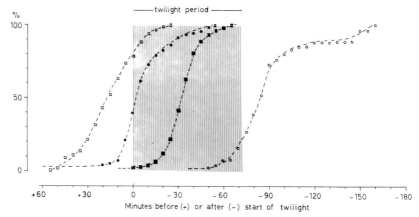

Fig. 12. Accumulated percentage of activity starts and ends of nocturnal and diurnal bullheads, respectively, in relation to the twilight periods. ■,□ = activity starts for nocturnal (n=902) and for diurnal (n=185) animals respectively. ●,○ = activity ends for nocturnal (n=902) and diurnal (n=214) animals.

ends take place before start of dawn and after end of dusk, respectively. Corresponding data from nocturnal bullheads (n=902) show that only about 2 percent of all activity onsets take place before dusk, and 40 percent of all activity ends before dawn (Fig. 12).

Firstly this means that 'normal' nocturnal animals have an activity time of about 11.9 h, diurnal ones of 13.9 h. Secondly, and more importantly, while very little indication of an endogenous timing component exists from the rhythmic behaviour of nocturnal bullheads, the spontaneous activity onsets in diurnal bullheads (and perhaps ends) indicate a circadian component. These differences are not related to the prevailing light conditions, but to diurnal or nocturnal phasing as such.

Feeding Activity of Nocturnal and of Diurnal Bullheads

Feeding activity of nocturnal as well as diurnal fish takes place mainly during dusk and before dawn (Fig. 13).

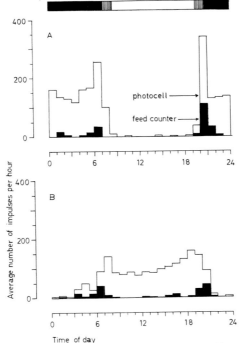

Fig. 13. The relation between locomotor and feeding activity in nocturnal (A) and diurnal (B) brown bullheads.

However, diurnal fish which mostly have some activity also during the dark period, feed to some degree also at night. Only about 33 percent of the feeding attempts of diurnal fish take place during the day (data from feeding activity

in diurnal animals are from two five-day periods only).
Thus the differences between nocturnal and diurnal bullheads are less pronounced as regards feeding than as regards locomotion.

*Endogenous Components in the Diel Activity Rhythm of
Brown Bullheads**

When brown bullheads are subjected to constant dark or
constant light, they become apparently arhythmic (Eriksson
and v. Veen, in prep.). It was suggested that the apparent
arhythmicity might depend on interactions between circadian
component(s) and variations in phototactic sensitivity.
In order to weaken the effect of such interactions without
giving experimental animals information about time of day,
we subjected brown bullheads to repeated 'dark pulses'
(1 lx) of 15 min duration each hour, against a background
light intensity of 400 lx.

Bullheads subjected to dark pulses were active during
dark periods only. However, activity was not equally distributed over the 24 h period, but occurred in a clearly
circadian manner for many days (Fig. 14). Frequency analysis indicated a circadian period of about 23 h. Although
endogenous circadian rhythms can be demonstrated in brown
bullheads under very special conditions only, they still
may have an important function in natural zeitgeber conditions. This is indicated by some experiments in heavy
water (Eriksson, in press). When a brown bullhead in LD
12:12 is subjected to 20 percent heavy water, its diel
rhythm is shifted in relation to the light-dark cycle.
The beginning of activity is later in the evening and its
end later in the morning; i.e. as would be expected of
synchronized endogenous rhythms reacting in relation to
the influence of heavy water.

In bullheads there is a direct effect of light on the
amount of activity during the diurnal phase of the 24 h
period. In addition there is an indirect effect of light
intensity on the phasing of the main activity pattern.
Feeding activity is apparently not shifted as completely
as locomotor activity, chiefly because feeding in caged
fish generally occurs around the twilight times, when
nocturnal as well as diurnal bullheads are active. The
way in which the change from nocturnalism to diurnalism
takes place, either as a fast transient process with one
nocturnal activity band running free and locking into the
new phase in relation to the light-dark cycle, or as a
slower process with nocturnal and diurnal patterns existing in parallel for long periods, is in complete agreement

*Footnote: This part is based on unpublished data which will be dis-
cussed in detail elsewhere (Eriksson and v. Veen, in prep.).

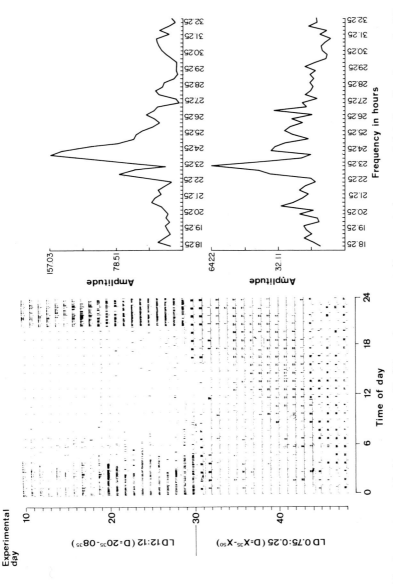

Fig. 14. Locomotor activity records (left) and corresponding periodograms (right) of a brown bullhead, exposed to LD 12:12 (A) and then to LD 0.75:0.25 (B). Data from Eriksson and van Veen, in preparation.

with what has been found in biphasic fish in natural light
assays at the Arctic circle (Müller, 1970a, b; Andreasson,
1973). The threshold light intensity for inversion within
a reasonable period of time in brown bullheads is very low,
about 1 lx. This would make it less probable for brown
bullheads to become diurnal in natural light. Even at
Messaure, light intensity is only rarely below 10 lx at
noon during midwinter, and here brown bullheads remain
nocturnal throughout the year. (Eriksson and Müller, unpublished data). The threshold for an inversion in burbots
(*Lota lota* L.) and sculpins (*Cottus poecilopus* Haeckel
and *C. gobio* L.) seems to lie somewhere between 100 and
10 lx (Müller, 1970a, b) and an inversion occurs in these
species.

Brown bullheads have an endogenous (circadian) rhythm,
difficult to establish in constant light conditions, but
possibly strongly affecting the activity of synchronized
fish.

There is no basis for regarding biphasic fish as very
special animals; the existence of a dual phasing capacity
of the diel activity rhythm is now well established also
in other animals. A change from nocturnalism in the summer
to diurnalism in the winter is typical for microtine voles
(Erkinaro, 1961, 1969), and they are clearly circadian in
constant conditions (Lehman, 1976). Many diurnal birds,
in which the expression of an endogenous rhythm is strong,
migrate at night. During the migratory phase nocturnal
and diurnal activity exist in parallel in a biphasic way
(see e.g. Gwinner, 1975b, fig. 28).

Conclusions

Brown trout and brown bullheads have a dual phasing
capacity of their diel locomotor activity patterns. Such
a capacity has now been clearly demonstrated in a number
of fish species and is indicated in several others, indicating that the phenomenon may be widespread among fish.

In the laboratory, a switch from one type of phasing
to another can be induced by changes in the prevailing
light conditions. Other cues have not been tested, and
nothing is known about the appearance of differences in
the phasing of individual fish in the field. However, it
is tempting to hypothesize that the dual phasing capacity,
using dawn and dusk as basic time cues, will give fish
flexibility in fulfilling their actual ecological needs
in a semi-opportunistic way. Systematic studies of how
biotic factors influence the phasing of biphasic fish
are most urgently required.

In addition, with the development of still more useful
tools for the study of rhythmic activity of fish in the
laboratory, such studies may add further valuable information on the physiological basis of circadian rhythms
in general (cf. Eriksson, 1973, 1975).

References

Andreasson, S. (1969). *Oikos* **20**, 78-94.
Andreasson, S. (1973). *Oikos* **24**, 16-23.
Aschoff, J. (1969). *Oecologia* **3**, 125-165.
Aschoff, J. and Wever, R. (1962). *J. f. Ornithologie* **103**, 1-27.
Byrne, J.E. (1971). *Can. J. Zool.* **49**, 1155-1158.
Chaston, I. (1968). *J. Fish. Res. Bd. Canada* **25**, 1285-1289.
Chaston, I. (1969). *J. Fish. Res. Bd. Canada* **26**, 2165-2171.
Daan, S. (1976). *Ibis* **118**, 223-236.
Daan, S. and Aschoff, J. (1975). *Oecologia* **18**, 269-316.
Eriksson, L-O. (1973). *Aquilo Ser. Zool.* **14**, 68-79.
Eriksson, L-O. (1975). Ph.D. Thesis, University of Umeå, Sweden.
Eriksson, L-O. (1977). *Experentia,* in press.
Eriksson, L-O. and van Veen, T. (in prep.). Circadian rhythms and phototactic behaviour in the brown bullhead, *Ictalurus nebulosus* (Teleostei): - Evidence for an endogenous rhythm in feeding-, locomotor-, and reaction-time behaviour.
Eriksson, L-O., Bauer, J., Habersetzer, J. and Hofman, L. (in prep.). A simple apparatus for the study of feeding activity of fish.
Erkinaro, E. (1961). *Oikos* **12**, 157-163.
Erkinaro, E. (1969). *Aquilo Ser. Zool.* **8**, 1-31.
Gwinner, E. (1975a). *J. comp. Physiol.* **103**, 315-328.
Gwinner, E. (1975b). *In: Avian Biology,* Vol. V, pp. 221-285. Academic Press, New York.
Harden Jones, F.R. (1956). *J. Exp. Biol.* **33**, 271-281.
Hoar, W.S. (1942). *J. Fish. Res. Bd. Canada* **6**, 90-101.
Kroneld, R. (1976). *Physiol. Zool.* **49**, 49-55.
Lehman, U. (1976). *Oecologia* **23**, 185-199.
Müller, K. (1969). *Aquilo Ser. Zool.* **8**, 50-62.
Müller, K. (1970a). *Oikos Suppl.* **13**, 122-129.
Müller, K. (1970b). *Oikos Suppl.* **13**, 108-121.
Müller, K. (1973). *J. comp. Physiol.* **84**, 357-359.
Penaz, M. (1975). *Zoologicke Listy* **24**, 263-276.
Pittendrigh, C.S. and Daan, S. (1976). *J. comp. Physiol.* **106**, 223-252.
Pohl, H. (1972). *Naturwissenschaften* **59**, 518.
Richardson, M.E. and McCleave, J.D. (1974). *Biol. Bull.* **47**, 422-432.
Snedecor, G.W. and Cochran, G.C. (1967). *Statistical Methods.* Iowa State Univ. Press, Ames, Iowa.
Swift, D.R. (1962). *Hydrobiologia* **20**, 241-247.
Swift, D.R. (1964). *J. Fish. Res. Bd. Canada* **21**, 133-138.
Varanelli, C.C. and McCleave, J.D. (1974). *Anim. Beh.* **22**, 178-186.
Westin, L. (1971). *Rep Inst. Freshw. Res. Drottningholm* **51**, 184-196.
Wikgren, B-J. (1955). *Mem. Soc. Fauna et Flora Fennica* **31**, 91-95.

THE FLEXIBILITY OF THE CIRCADIAN SYSTEM OF FISH AT DIFFERENT LATITUDES

KARL MÜLLER*

*University of Umeå,
Department of Ecological Zoology,
S-901 87 Umeå, Sweden.*

Introduction

Many species of the European fish fauna have a wide geographical distribution. The species from the River Kaltisjokk (Arctic Circle, 66°42'N, 20°25'E) (*Salmo trutta, Lota lota, Cottus poecilopus* and *Phoxinus phoxinus*) have been found between 45°N and 71°N, including the British Isles. These fish species are relatively insensitive to differences in chemical and physical conditions within their distribution area. In particular they are exposed to great variations in the daily and seasonal changes of the light-dark cycle and therefore they must be extremely flexible in their reactions to the photoperiodic conditions at different latitudes. From the behavioural point of view, the light-dark regime is the dominant factor affecting the activity phase of an organism.

In the following I will make a survey of diel and seasonal fish locomotor patterns at different latitudes between 47°N and the Arctic Circle, based on work of several authors (Table 1).

Results

The diel activity of fish is normally synchronized with the alternation of light and dark in the 24-h period. The basic attributes of such diel periodicity have been demonstrated for various organisms. But the diel activity at high latitudes can be very different from that at lower latitudes, which is related to the extreme light conditions at the summer and winter solstices and to the great seasonal differences in the photoperiod and light inten-

*Footnote: Supported by the Swedish Natural Research Council, the Max-Planck-Gesellschaft zur Förderung der Wissenschaften and the Deutsche Forschungsgemeinschaft.

TABLE 1

Investigations on fish rhythmicity in Central and Northern Europe

Locality	Species	Author
1. Mondsee 47°48'N, 13°26'E	*Lota lota* *Phoxinus phoxinus* *Salmo trutta*	Müller Müller Müller and Eriksson
2. Erling-Andechs 47°58'N, 11°30'E	*Ictalurus nebulosus*	Eriksson
3. Dietershausen 50°30'N, 9°52'E	*Cottus gobio* *Salmo trutta*	Müller Eriksson
4. Berlin 52°30'N, 13°25'E	*Perca fluviatilis* *Tinca tinca* *Scardinius erythroph-* *thalmus*	Siegmund Siegmund Siegmund
5. Hamburg 53°N30'N, 10°00E	*Phoxinus phoxinus* *Lampetra fluviatilis* *Lota lota* *Anguilla anguilla*	Wehrmann Tesch Koops Tesch
6. Lund 55°35'N, 13°30'E	*Cottus gobio* *C. poecilopus* *Lota lota*	Andreasson Andreasson Müller
7. Turku 60°25'N, 22°30'E	*Lota lota*	Wikgren
8. Trondheim 63°14'N, 10°26'E	*Lota lota*	Solem
9. Hölle 64°08'N, 20°15'E	*Salmo trutta* *Lampetra fluviatilis*	Eriksson Sjöberg
10. Rickleån 64°05'N, 20°56'E	*Lampetra fluviatilis* *Lampetra fluviatilis*	Fogelin Sjöberg
11. Messaure 66°42'N, 20°25'E	*Salmo salar* *Salmo trutta fario* *Salmo trutta trutta* *Salvelinus fontinalis* *Coregonus lavaretus* *Lota lota* *Cottus poecilopus* *Phoxinus phoxinus* *Perca fluviatilis* *Cottus gobio* and *C. poecilopus*	Eriksson Müller Eriksson Eriksson Müller Müller Müller Müller Eriksson Andreasson
12. Abisko 68°21'N, 18°49'E	*Cottus poecilopus* *Phoxinus phoxinus*	Müller Müller

sities in the north. This has been shown in many investigations (Erkinaro, 1961, 1969; Peiponen, 1962; Nyholm,
1965; Müller, 1968, 1969, 1970, 1973, 1976; Eriksson,
1973, 1975) primarily concentrated to the area around
the Arctic Circle. Fish species (e.g. *Salmo trutta, S.
salar, Coregonus lavaretus, Phoxinus phoxinus, Cottus
poecilopus, Lota lota*) in the subarctic region can have
aperiodic rhythmic patterns around the summer solstices:
they have no synchronized alternation between activity
and rest from May to the beginning of August (at Messaure
20 km north on the Arctic Circle). Another type of reaction to the extreme light conditions around midsummer is
a 'free-running' circadian rhythm which was observed in
four species at Messaure: *Phoxinus phoxinus* (Müller,
1968), *Salmo trutta* (Müller, 1969), *Cottus poecilopus*
(Müller, 1970) and *Salvelinus fontinalis* (Eriksson, 1972).
Such phenomena of arhythmicity or desynchronization have
not been observed at lower latitudes.
     Another important factor is the initiation of activity
of an organism. Increasing or decreasing light intensity
function as signals for onset of activity for diurnal and
nocturnal species, respectively. The strength of these
zeitgebers differs during the year and depends upon the
duration of twilight. Under the extreme light conditions
at summer and winter solstices the strength of the zeitgeber is reduced for both diurnal and nocturnal organisms.
The effect of the zeitgeber on circadian rhythms disappears more rapidly in conditions of almost perpetual illumination, than in perpetual darkness (Bruce, 1960;
Aschoff, 1963).

Investigations of Night-Active Fishes

The Burbot (Lota lota L.)   The burbot (*Lota lota*) has
a range extending from the Alps and Pyrenees to Northern
Europe and European Russia. The burbot occurs in running
water, lakes, estuaries and coastal areas of the Baltic
and North Sea. It undertakes regular migrations from
brackish water into rivers as has been shown by Müller
(1960), Koops (1959, 1960) and Tesch (1967).
     After learning how to maintain burbot under experi-
mental conditions and how to measure the activity of the
species continuously over several years I was able to
describe the seasonal system of the phase inversion
(Müller, 1969, 1970, 1973). At the beginning of September
the burbot are clearly night-active (Fig. 1). In mid-
September a peak of activity develops at midday, but the
overall pattern of night-activity persists (Fig. 2).
At the beginning of October a clear change to day-activity
occurs (Fig. 3).
     For eight simultaneously investigated immature burbot
the phase inversion began in the second half of September

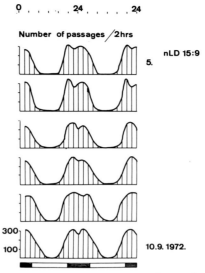

Fig. 1. The locomotor activity of the Burbot (*Lota lota*) from September 5, to 10, 1972. Dark areas in the bar mean night time, white area day time.

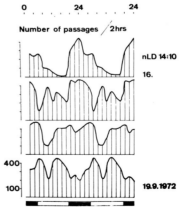

Fig. 2. The locomotor activity of the Burbot (*Lota lota*) from September 16, to 19, 1972. Dark areas in the bar mean night time, white area day time.

and was completed during October. The reverse process started at the end of January or the beginning of February and was completed at the beginning of March (Fig. 4). Sexually mature males and females completed their phase inversion earlier or at the end of January and the beginning of February. These adults were markedly night-active ap-

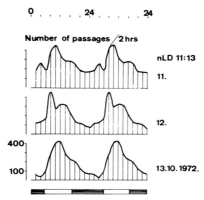

Fig. 3. The locomotor activity of the Burbot (*Lota lota*) from October 11, to 13, 1972. Dark areas in the bar mean night time, white area day time.

proximately 10-15 days before the spawning time (Müller and Österdahl, 1970).

The activity patterns around the year have been observed at different latitudes: Messaure ($66°N$), Lund ($55°N$), Dietershausen ($51°N$) and Mondsee ($47°N$).

The results are demonstrated in Fig. 5. At lower latitudes the burbot is always clearly night-active: north of

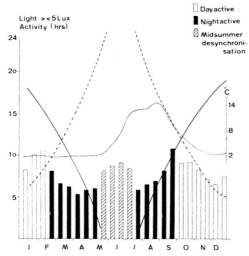

Fig. 4. Duration of activity and phase-position of eight burbots (*Lota lota*) in relation to the length of day and night during the year 1970 in Messaure. Light >.5 lx = area below dotted lines: light < 5 lx = are below full lines. Water temperature in $°C$.

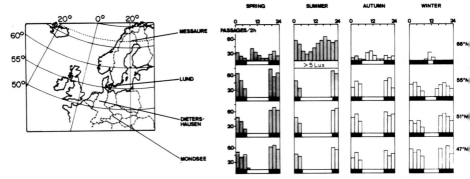

Fig. 5. The daily distribution of locomotor activity of the Burbot (*Lota lota*) in spring, summer, autumn and winter at different latitudes.

the Arctic Circle the fish show the change of the phase position (spring, autumn), an arhythmic locomotor pattern in midsummer, and day-activity in midwinter. Thus the diel and seasonal rhythmic behaviour of the burbot differs according to the latitude. In the southern parts of its distribution area night-activity is maintained, while a phase inversion occurs at northern latitudes during spring and autumn. In the autumn and spring this coincides in adult fishes, with the upstream and downstream migration of the coastal population (Koops, 1959, 1960) in the River Elbe and in the North Swedish coastal streams Ricklean (Bengtsson 1973) and Angerån (unpubl. data, 1977). The region of transitions between clearly synchronized activity patterns and phase inversion conditions are situated around 60°N (Wikgren, 1955) and 64°N (Bengtsson, 1973).

Investigations carried out at Messaure over a whole year in constant conditions with a light-dark change of 12:12 (D = 0.1 lx, L = 25 lx) show that the burbot even under these conditions have the phase inversion at the same time as under natural light-dark conditions (Fig. 6). This can be explained by the phase inversion mechanism being bound to the endogenous seasonal activity pattern.

The Sculpins (Cottus poecilopus Heckel and C. gobio L.)
Both species have been thoroughly investigated by Andreasson (1969, 1973), Andreasson and Müller (1969) and Müller (1970). The significance of falling (or rising) light intensity for phase inversion in these species has been described by Müller (1970).

C. poecilopus was transferred from Messaure (66°N) to Lund (55°N) in November after completed phase-inversion. These specimens maintained the phase-position acquired at Messaure which could have been predicted according to the photoperiodic conditions in Lund. In March-April a shift to summer night activity occurred (Andreasson and Müller,

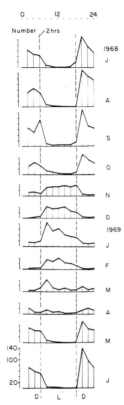

Fig. 6. The locomotor activity of the Burbot (*Lota lota*) under artificial light-dark (12:12) conditions from July 1968 to June 1969 in
the Messaure laboratory. Dotted lines = light on (0600 h) and light
off (1800 h).

1969). *C. poecilopus* transferred from Lund to Messaure in
August showed an inversion during September-October, at
the same time as the native fish from the River Kaltisjokk
near Messaure (Müller, 1970). The other sculpin species,
Cottus gobio, showed a similar phase-inversion at high
northern latitudes (Andreasson, 1973).

In 1973 and 1974 I carried out investigations on *Cottus
gobio* in Central Europe (Dietershausen, 51°N), on fishes
exposed to natural light conditions. The experimental
groups were placed at different distances from a window
directed eastwards in order to give varying light intensities.

One group (four replicates) was exposed to approximately
500 lx as maximum light intensity during mid-winter. These
specimens were completely night-active throughout the

experiment (Fig. 7).

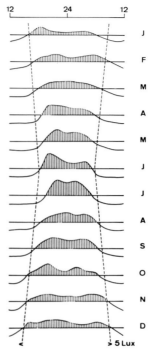

Fig. 7. The locomotor activity (monthly means) of the sculpin (*Cottus gobio*) at Dietershausen (51^0N, West Germany) from November 1973 to October 1974. Ordinate: Percent deviation from the 24 h mean. Abscissa: Daytime in hours.

In the second group (four replicates) with the winter light intensity never exceeding 200 lx the beginning of activity in the afternoon shifted forward during November and December (Fig. 8). For the last group (four replicates) the winter light-intensity never rose above 20 lx. Under these conditions a gradual inversion towards day activity began in November and went on to the end of December. In January they assumed a crepuscular activity with peaks at dawn and dusk, but at the end of January they were clearly night-active again (Fig. 8).

Attempts to induce the phase inversion at other times of the year than September/October and February/March were unsuccessful. This can be explained only by the phase inversion mechanism being bound to the endogenous seasonal activity pattern.

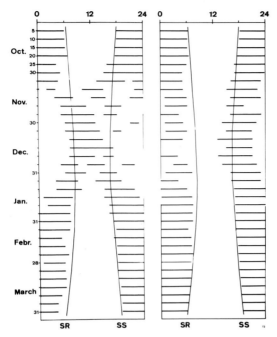

Fig. 8. The locomotor activity and phase-position (five day means)
of the sculpin (*Cottus gobio*) in Dietershausen from October 1973
to March 1974. At right: Daily maximal light 200 lx, at left: Daily
maximal light 20 lx. SR, sunrise: SS, sunset.

Investigations of Day-Active Fishes

Siegmund (1969) conducted experiments with the day-
active fishes, *Scardinius erythrophthalmus* and *Perca
fluviatilis* (Berlin, 52°N). They were clearly day-active
throughout the year, but the activity peaks changed according to the length of the photoperiod. The amount of activity was proportional to the water temperature.

The Minnow (Phoxinus phoxinus L.)   This species has been
investigated at various latitudes - 47°N (Mondsee/Austria),
51°N (Dietershausen/West-Germany), 55°N (Lund/Sweden and
66°N (Messaure/Swedish Lapland) and the results are summarized in Fig. 9.

In Central Europe and Southern Sweden the minnow is
clearly day-active throughout the year. But during the
subarctic summer conditions, when light intensity is always above 5 lx (at Messaure May 20 to July 20) the minnow has an arhythmic activity. The same activity pattern
applies for all genuine day-active fish species examined
at Messaure, viz. *Coregonus lavaretus*, *Perca fluviatilis*,

Salmo trutta, S. salar, salvelinus alpinus and *fontinalis*.
The minnow can be desynchronized or reveal a 'free-running'
circadian periodicity during that time, when the sun is
constantly above the horizon. At other times of the year
the species is day-active.

The flexibility of the circadian rhythm in the minnow
is expressed as a desynchronization of the activity in
the north resulting from the extreme reduction of the
strength of the zeitgeber at midsummer.

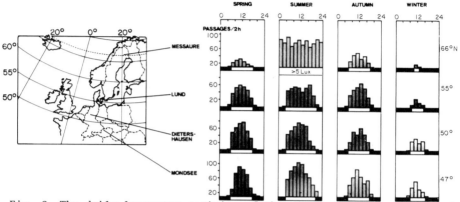

Fig. 9. The daily locomotor activity of the Minnow (*Phoxinus phoxinus*)
in spring, summer, autumn and winter at different latitudes.

The Flexibility of Phase-Position at Spawning Time of Fishes

Enequist (1937) observed for the first time that ord-
inary locomotor activity and spawning activity of the
river lamprey (*Lampetra fluviatilis* L.) occurred at different times of the day in the lower part of the River
Luleälv (65°N). Wikgren (1953) confirmed the results of
Enequist. Müller and Österdahl (1970), and Müller (1970)
found the same phenomenon in *Lota lota* and *Cottus poecilopus* respectively. Sjöberg (1977) studied the activity of
the river lamprey during the spawning time, and he demonstrated very clearly the 'locomotor peak' during the night
and the 'spawning-peak' during the day. Thus, the spawning activity in many species is clearly separated from
the locomotor activity, perhaps comparable to the migratory restlessness of birds. Both activities are always
synchronized with the 24 h period.

The Locomotor Activity of *Barbus partipentazona* Transferred to High Northern Latitudes

The activity of *Barbus partipentazona*, transferred
from Singapore to Messaure, was studied under natural
light-dark conditions and with constant water temperature

(25°C) at the Arctic Circle. It was synchronized with
the 24 h period during the whole year. The locomotor activity was rather low during the first three months of the
year, but increased rapidly in April, culminated in June
and declined steadily after that. In midsummer the duration of activity never exceeded 14-15 h in this day-active
species (Müller and Figala, 1972).

This experiment was performed during one year only.
Perhaps the activity adjusts to the light-dark changes
over a longer period, analogous with the pattern observed
in native day-active fishes. On the other hand, the diel
periodicity in this tropical fish may be genetically established and in this case it will be maintained ever under
extreme conditions of illumination.

Discussion

The investigations reported here show that it is not
possible to generalize on the activity behaviour of fish.
The circadian system or the 'circadian clock' has a high
degree of adjustability or flexibility to the environmental conditions.

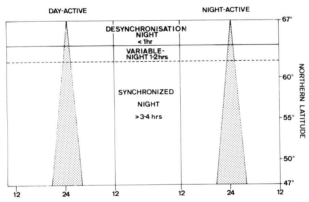

Fig. 10. Synchronization and desynchronization of day- and night-
active fishes at midsummer at different latitudes. Fishes are always
synchronized with the 24 h period if night length is more than 3-4 h.
In areas with midsummer night duration between 1-2 h synchronization
is unstable or weak. With night duration shorter than 1 h day- and
night-active fishes are desynchronized. Dotted areas: duration of
the night at different latitudes.

At the Cold Spring Harbor Symposium, Pittendrigh (1960)
emphasized the necessity of more research in Arctic and
Antarctic regions, where the daily solar rhythm is absent
or reduced during some time of the year.

The favourable geographical position of northern Fenno-
scandia, which permits long-term, continuous research to
be carried out on organisms in their natural environment,
has enabled us to extend knowledge of the function of circadian systems.

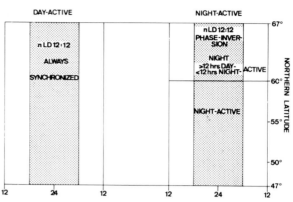

Fig. 11. Aspect of synchronization in fish at different latitudes
during autumn and spring. Day-active fishes are always synchronized
with the 24 h period: night-active fishes too, up to about 60°N latitude. At higher northern latitudes ($>60^{\circ}$N) night-active fishes shift
their activity into daytime after the autumn equinox, and they invert
their activity from day to night in the early spring (February - March).
Dotted areas = Duration of the night.

The results of the work up to now are summarized in
Figs. 10-12. Up to 60°N all fishes are synchronized during
the summer (Fig. 10). Further north the strength of the
zeitgeber is so much reduced that it can cause desynchronization of the regular alternation of activity and rest.
Above 64°N the summer nights are reduced to 1 h or less
and both day-active and night-active fish as well as other
aquatic organisms show an arhythmic or desynchronized
locomotor behaviour. The effect of the zeitgeber on circadian rhythm disappears more rapidly at an extreme reduction of darkness, than at an extreme reduction of light.
Locomotor activity in spring and autumn is presented
in Fig. 11. At both seasons natural light-dark conditions
are 12:12. At lower latitudes most animals and plants show
a conspicuous synchronization at these times. This is also
the case for day-active fishes at high latitudes, but for
night-active fishes it is true only up to 60°N. With shortening of the day to less than 12 h, combined with a reduction of light intensity, night-active fishes change their
phase position from night to day. The opposite holds true
in the spring.

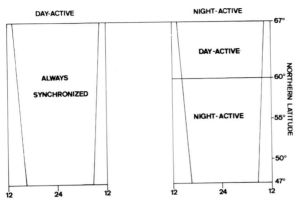

Fig. 12. The winter aspect of phase-position in day- and night-active fishes at different latitudes. All fishes are synchronized with the 24 h period at all latitudes. The 'night-active' fishes are night-active up to about the 60°N latitude, but further north they are day-active. Dotted area: duration of the night at different latitudes.

In winter (Fig. 12) the phase-shifted species are day-active. At lower latitudes the maxima of activity of night-active fish species occur around dawn and dusk.

The adaption to the different environmental conditions and structures at different latitudes must be seen as of positive value for the fish's survival, and the flexibility in rhythmic patterns expressed in a disbanding of the temporally regulated changing of activity and rest would be selectively advantageous for the species, for example in utilizing the rich summer food resources at high northern latitudes optimally.

References

Andreasson, S. (1969). *Oikos* **20**, 78-94.
Andreasson, S. (1973). *Oikos* **24**, 16-23.
Andreasson, S. and Müller, K. (1969). *Oikos* **20**, 171-174.
Aschoff, J. (1963). *Ann.Rev.Physiol.* **25**, 581-600.
Bengtsson, B. (1973). Ekologiska studier på Lake (*Lota lota* L.). University of Umeå Dept. of Ecological Zoology 1-30.
Bruce, V.G. (1960). *Cold Spring Harbor Symp.Quant.Biol.* **25**, 29-48.
Chaston, I. (1968). *J. Fish. Res. Bd Canada* **25**, 1285-1289.
Enequist, P. (1937). *Ark. Zool. Uppsala* **29**, 1-20.
Eriksson, L.O. (1972). *Fauna och Flora* **67**, 233-234.
Eriksson, L.O. (1973). *Aquilo, Ser. Zool.* **14**, 68-79.
Eriksson, L.O. (1975). Ph.D. thesis, University of Umea, 1-43.
Erkinaro, E. (1961). *Oikos* **12**, 157-163.
Erkinaro, E. (1969). *Aquilo, Ser. Zool.* **8**, 1-30.

104   K. MÜLLER

Erkinaro, E. (1969). *Z. vergl. Physiol.* **64**, 407-410.
Fogelin, P. (1972). *Rapp. Rickleå Fältst.* **20**, 1-11.
Koops, H. (1959). *Mitt. Inst. Fischereibiologie Hamburg* **9**, 1-60.
Koops, H. (1960). *Mitt. Inst. Fischereibiologie Hamburg* **10**, 43-55.
Müller, K. (1968). *Naturwissenschaften* **55**, 141.
Müller, K. (1969). *Aquilo, Ser. Zool.* **8**, 50-62.
Müller, K. (1969). *Experientia* **25**, 1268.
Müller, K. (1970). *Oikos, Suppl.* **13**, 108-121.
Müller, K. (1970). *Oikos, Suppl.* **13**, 122-129.
Müller, K. (1971). *Natur. und Museum* **101**, 146-154.
Müller, K. (1973). *Aquilo, Ser. Zool.* **14**, 1-18.
Müller, K. (1973). *J. comp. Physiol.* **84**, 357-359.
Müller, K. (1976). *Arch. Fisch. Wiss.* **27**, 121-132.
Müller, W. (1960). *Zeitschr. f. Fischereiwiss.* **9**, 1-72.
Müller, K. and Österdahl, L. (1970). *Oikos, Suppl.* **13**, 130-133.
Müller, K. and Figala, J. (1972). *Aquilo, Ser. Zool.* **13**, 13-20.
Nyholm, E.S. (1965). *Ann. Zool. Fenn.* **2**, 77-123.
Peiponen, V.A. (1962). *Arch. Soc. Vanamo* **17**, 171-178.
Peiponen, V.A. (1970). *In: Ecology of subarctic region.* Proc. Helsinki
    Symp. Unesco pp. 281-287.
Pittendrigh, C.S. (1960). *Cold Spring Harbor Symp. Quant. Biol.* **25**,
    159-184.
Siegmund, R. (1969). *Biol. Zbl.* **88**, 295-312.
Sjöberg, K. (1974). *Zool. Revy* **36**, 41-48.
Sjöberg, K. (1977). Locomotor activity of the river lamprey (*Lampetra
    fluviatilis* L.) during the spawning season. (In.press).
Solem, J.O. (1973). *Oikos* **24**, 328-330.
Tesch, F.W. (1967). *Helgoländer Wiss. Meeresunters.* **16**, 92-111.
Wehrmann, L. (1975). Ph. D. thesis, University of Hamburg, 1-89.
Wikgren, B.J. (1953). *Mem. Soc. Fauna et Flora Fenn.* **29**, 24-27.
Wikgren, B.J. (1955). *Mem. Soc. Fauna et Flora Fenn.* **31**, 91-97.

ANNUAL CYCLES IN THE LIGHT ENVIRONMENTS AND VISUAL MECHANISMS OF FISHES

W.R.A. MUNTZ* AND A.W. WAINWRIGHT[+]

*Department of Biology,
University of Stirling, Stirling, FK9 4LA
Scotland.

[+]Centre for Research on Perception and Cognition,
University of Sussex, Brighton BN1 9QY,
England.

Introduction

The retinas of many fishes, particularly freshwater species, have two extractable visual pigments, one based on retinol (A_1-based) and the other on 3-dehydroretinol (A_2-based). Since in such paired-pigment species the A_1-based pigment absorbs light at shorter wavelengths than the A_2-based pigment, the total absorbance, and presumably therefore the animal's spectral sensitivity (at least under scotopic conditions), will depend on the relative proportion of the two types of pigment. In 1961, Dartnall *et al.* demonstrated that in the rudd (*Scardinius erythrophthalmus*) this proportion is not fixed, but depends on the lighting conditions under which the fish are kept: with long day-lengths the proportion of the A_1-based pigment increases, while with short daylengths the reverse occurs. Subsequent behavioural experiments have shown that, as we should expect, there is a concomitant change in the scotopic spectral sensitivity curve (Muntz and Northmore, 1973). It is now further known that not only can the relative proportion of the A_1- and A_2-based pigments of a number of other species, apart from the rudd, be altered by varying the light regime under which they are kept, but that several other factors, including hormones, age, retinal location sampled, and temperature, can also be correlated with changes in the proportion of the two pigment types (see Bridges, 1972; Crescitelli, 1972; Beatty, 1975; for reviews).

Since several of these factors, such as temperature and light, vary with the seasons, and since it has been shown for several species of fishes, obtained at different times of the year from their normal habitats, that changes in the pigments occur under natural conditions (Fig. 1), it is reasonable to suppose that the pigment changes are adaptations to seasonal variations in the quantity or

spectral quality of the light in the environment. Fig. 1
shows, however, that the time course of the changes varies

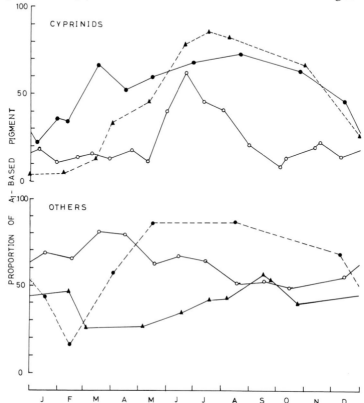

Fig. 1. Seasonal changes in visual pigments of freshwater fishes in
their natural habitats. The top half of the figure shows the results
for cyprinids, the bottom half for fishes of other groups. Where more
than one sample was taken at nearly the same time of year, the results
have been averaged. For the cyprinids the triangles and broken line
show the results for *Richardsonius balteatus balteatus* (Allen, 1971),
the empty circles and continuous line for *Scardinius erythrophthalmus*
(present paper), and the filled circles and continuous line for
Notemigonius crysoleucas boscii (Bridges, 1965). Bridges (1965) also
presents less complete data for *Notemigonius* from a second location
where the water was clearer: the results were similar except that at
any given time of year a greater proportion of A_1-based pigment was
present. In the lower half of the figure the filled circles and broken
line show results for *Balonesox belizanus* (Bridges, 1965), the empty
circles for *Lota lota* (Beatty, 1969), and the triangles for *Salmo
clarki* (Allen *et al.*, 1973). *Salmo clarki* from a shaded area of the
same stream showed much smaller seasonal variations, ranging from 5.74%
to 64.6% A_1-based pigment over the year.

widely in different species. Either, therefore, the light
environments of the various species alter in different
ways over the year, or the adaptive significance of the
pigment changes is not the same in all cases. As a first
step towards understanding the significance of the seasonal visual pigment changes of freshwater paired-pigment
fishes, it is therefore necessary to measure the light in
a lake that actually contains such a species, and at the
same time to catch samples of fish at regular intervals
and characterise their visual pigments. The present paper
reports the results of such a study, made on a small lake
in Sussex that contains a population of rudd (*Scardinius
erythrophthalmus*).

Methods

Light Measurements

The light measurements were made with the apparatus
shown schematically in Fig. 2. The detector (A) was PIN
silicon photodiode with a 2 cm diameter receptor surface,
integrated with an FET operational amplifier (UDT-500,
United Detector Technology Inc.), and mounted below a
rotating disc (B) containing five Balzars B-40, 32 mm diameter, interference filters.

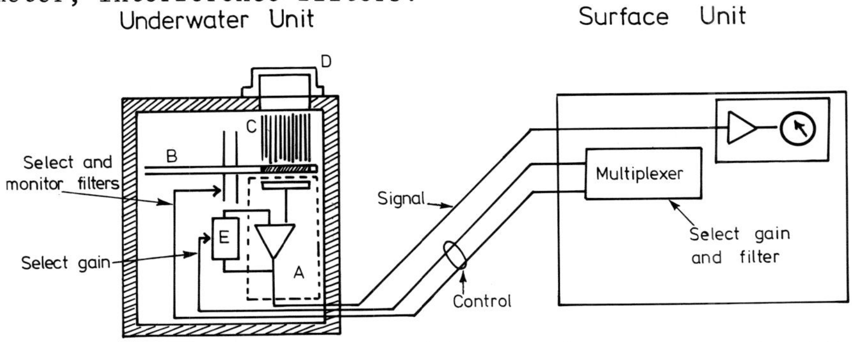

Fig. 2. Schematic diagram of the light measuring apparatus. A: silicon
PIN photodiode and FET operational amplifier. B: disc with interference
filter. C: collimation tubes. D: removable cosine collector. E: feedback resistors.

The spectral characteristics of the filters are summarised
in Table 1. The light entered the apparatus through a 2.5
cm hole, sealed with a clear Perspex window, and was col-
limated by an array of drinking straws painted matt black
(C) before reaching the interference filters. A removable
opal Perspex cosine collector (D) with an O-ring seal could
be mounted outside the clear window. With the opal Perspex
in place the instrument was used to measure irradiance,
and its angular response in this mode is shown in Fig. 3.
The instrument was also used without the opal to measure

radiance; the angular response in this case also is shown in Fig. 3.

TABLE 1

Interference filter characteristics

λmax (nm)	387.5	451.5	510.0	578.0	631.5
Half maximum bandwidth (nm)	7.5	11.5	6.5	7.5	9.0

The different interference filters were inserted into the beam by stepping the disc (B) with a solenoid, and the position of the disc was monitored on the surface by signals from micro-switches activated by studs on the disc's axle. The gain of the instrument was controlled by different feedback resistors (E), which were inserted by relays activated from the surface. For most of the measurements the instrument was mounted in a frame with three horizontal arms, so that it could be suspended facing either upwards or downwards to collect the downwelling or upwelling light respectively. Some radiance measurements were also made with a mounting that allowed the instrument to be rotated round the vertical plane in 45° steps.

The instrument was calibrated in relative terms using a Hilger-Schwartz F 17 thermopile, and in absolute terms using a solution of visual pigment as a chemical actinometer. A beam of light from a stable light source, consisting of a tungsten-iodine bulb run from a stabilised supply, was focussed on to the receiving window of the thermopile, and the resulting voltage measured on a Solartron digital voltmeter. The intensity of the light was measured in this way with various interference filters of known characteristics interposed in the beam. These results allowed an estimate to be made of the colour temperature of the lamp, using the tables given in Moon (1961) and Wyszecki and Stiles (1967). The absolute intensity of the light was then also estimated, by bleaching a solution of bovine visual pigment placed in a cuvette of known area at a fixed position in the light beam. The initial transmission of the solution, and the transmissions at various times following exposure to the light source with a 531 nm interference filter placed in the beam, were measured with a Pye-Unicam SP500 spectrophotometer. The solution was finally bleached completely, a photometric curve constructed, and the quanta sec^{-1} cm^{-2} entering the cuvette estimated using the formulae given in Dartnall (1968). Once the colour temperature and intensity at 531 nm of the light source were known, it was possible to determine the absolute amount of energy available at each wavelength.

By placing the light measuring instrument in front of the source, a direct calibration could then be made. This was done on two occasions, separated by six months, and the agreement between the two sets of data was good.

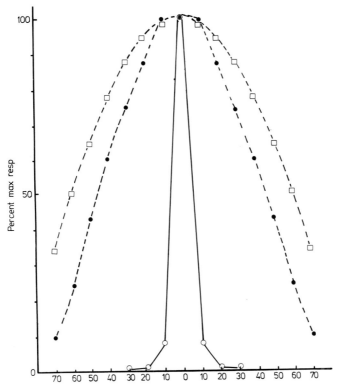

Fig. 3. Angular response of the apparatus with (filled circles and dashed line) and without (empty circles and continuous line) the cosine collector in place. The empty squares and dashed line show the cosine law.

The measurements were made from a 7 ft inflatable boat, to which was attached a framework supporting a 4 ft extensible boom, from which the instrument was suspended.

Pigment Extractions

Samples of fish were collected at regular intervals by seine netting, electro-fishing, or occasionally by rod and line. They were dark adapted overnight, killed by decapitation, and the retinas dissected out and deep-frozen for later analysis.

Separate visual pigment extracts were made from the

retinas of each individual fish, using conventional procedures as described, for example, in Muntz and Northmore
(1971). The retinas were washed several times in acid
buffer (pH 4.6) and extracted overnight in 3% digitonin
solution. Ten percent by volume of saturated sodium borate
solution and the same amount of 0.2 M hydroxylamine solution were added before measuring the spectral absorbance.
The results in the present paper refer in all cases to
difference spectra calculated from the spectral absorbance
of the solution before and after a 20-25 min. bleach with
tungsten light passed through a Wratten No. 15 filter.
The relative proportions of the A_1- and A_2-based pigment
were calculated from these difference spectra using the
computer program described in Muntz and Northmore (1971),
and assuming that the two pigments of this species are
$VP507_1$ and $VP535_2$ (Bridges and Yoshikami, 1970).

Doctor's Lake

The light measurements and fish collections were made
at Doctor's Lake, near Horsham, in Sussex. The lake, which
is surrounded by trees and located in a slight hollow, is
approximately 140 m long by 40 m wide at its widest point,
with its long axis extending in an east to west direction.
It is fed by two small streams at the east end with the
outflow over a weir at the west end. The eastern half of
the lake is shallow (0.6-0.9 m) and largely occupied by
rushes, whereas the western half is deeper (1.2-1.5 m).
Most of the light measurements were made in the western
half of the lake in the deeper water.

Results

Light Measurements

Irradiance Measurements With a few exceptions, irradiance
measurements (with the cosine collector in place) were
made at weekly intervals. On each occasion the measurements were made as nearly as possible at 1200 h GMT, for
both upwelling and downwelling light, at depth intervals
of 0.3 m down to 1.2 m. Variations in the density of the
cloud cover caused changes of up to 0.8 log units in the
light intensity at the surface of the lake. As far as
possible the measurements were made during periods when
the cloud conditions were constant but on occasions some
variation from this cause was unavoidable.
        Fig. 4 shows an example of how the downwelling light,
expressed as a proportion of the surface light, varied
with depth. As we should expect, when the intensity of
the light is expressed in logarithmic units the resulting
plots against depth are approximately straight lines, the
slopes of which give the downward attenuation coefficients
(K+) for the different wave-lengths. If the lines are ex-
trapolated back to zero depth they fail to intersect the
intensity axis at the 100% point due to the 'surface

effect' on the cosine collector (Tyler and Smith 1970).
Fig. 4 also shows how the upwelling light varied with
depth on this occasion.

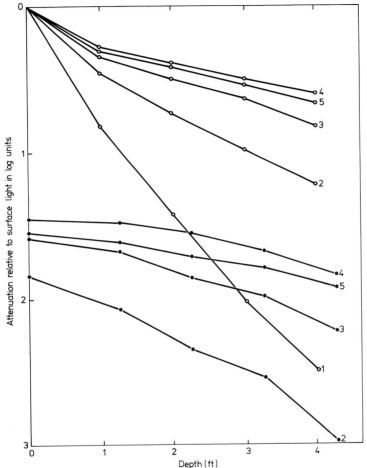

Fig. 4. Attenuation of downwelling (empty circles) and upwelling
(filled circles) irradiance with depth. Results from site 2 on 20th
September 1974. The results for the filter with λmax at 387.5 nm,
415.5 nm, 510.0 nm, 578 nm, and 631.5 nm are labelled 1 to 5 respectively.

   Fig. 5 shows some examples, for downwelling light, of
the percentage of the surface light remaining at a depth
of 0.6 m, at each wavelength. K+ can be estimated from
these data after allowing for the surface effect, and is
shown by the right hand axis. Fig. 5b shows similar spec-

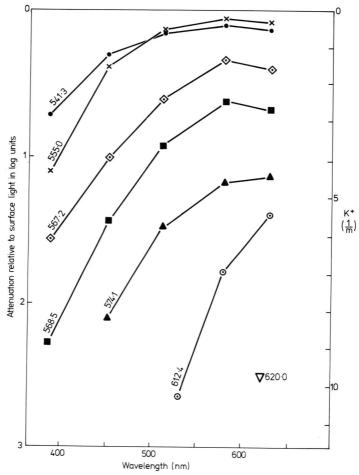

Fig. 5a (see legend opposite page)

tral curves for the upwelling light. There was much less upwelling light than downwelling light, and it was often impossible to make any measurements at 0.6 m. In this case therefore, the curves give the upwelling light at 0 m (that is, with the cosine collector just beneath the surface and pointing downwards) as a percentage of the downwelling light in air at the surface.

The examples shown in Fig. 5 cover the range of conditions that were encountered during the year. The curves for the downwelling light form a regular series, with the light having relatively less short wavelength energy as the attenuation became greater. The correlation between

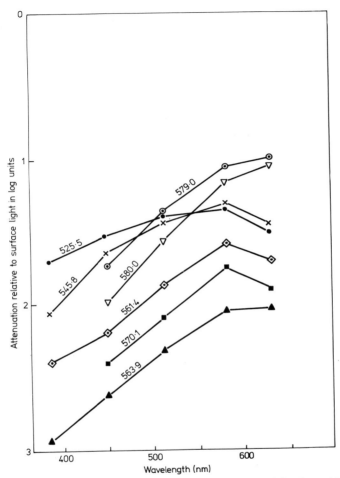

Fig. 5b Sample curves for different days showing (*a*) the attenuation of the downwelling irradiance at 0.6 m, and (*b*) the attenuation of the upwelling irradiance at 0 m. The values in all cases are relative to the downwelling irradiance at the surface, and the figures beside each curve give its centre of gravity (CG). The right hand (K+) ordinate in (*a*) was constructed assuming a 'surface effect' as in Fig. 4. The data were obtained on 3rd May 1974 (empty diamonds), 9th August 1974 (upright filled triangles), 11th October 1974 (crosses), 15th November 1974 (empty circles), 22nd November 1974 (inverted filled triangles), 7th February 1975 (filled circles) and 25th July 1975 (filled squares). On many days there was insufficient light per measurements to be made at all wavelengths, e.g. the 22nd November 1974, when downwelling measurements were only possible at 631.5 nm.

the spectral quality of the light and the degree of attenuation was not, however, perfect, as may be seen by comparing the results for 11th October 1974 and 7th February
1975. A similar difference between the spectral quality
of the light on these two days may also be seen in the
data for the upwelling light (Fig. 5b).

The curves for upwelling light shown in Fig. 5b also
form a regular series, with the exception of those obtained
on 9th August 1974 and 15th November 1974. On these two
days the water was heavily discoloured with suspended mud,
and the attenuation of the downwelling light was very high
(Fig. 5a). At O m there was however, more upwelling light
at long wavelengths than on even the clearest days, presumably because of the high degree of backscatter caused
by the suspended material. This did not apply to the short
wavelength light, which was heavily attenuated and unmeas-
ureable at 387 nm. The rate at which the upwelling light
decreased with depth was also very great under such conditions, so that although there was more long-wavelength
upwelling light at the surface, compared to clear days,
by the time a depth of about O.6 m was reached there was
less light at all wavelengths.

For the purpose of data compression the centres of gravity (CG) of the different spectral curves, such as those
shown in Fig. 5, were calculated as follows. For each set
of results the five readings were multiplied by the wavelengths at which they were obtained; these five products
were then summed, and divided by the sum of the readings.
CGs can be calculated in this way either for readings expressed in absolute terms, or for readings expressed as a
proportion of the surface light, and in either case a single figure representing the spectral characteristics of the
light is obtained. In view of the large number of light
measurements that were made, some form of data compression
was essential. It must however be remembered that the values of the CG depend to some degree on the spectral position of the interference filters used, and that on some
occasions no measurements could be made at the shortest
wavelengths and a reading of zero had therefore to be
assumed in calculating the CG. In spite of these defects
the CG provides a straightforward measure of the spectral
quality of the light, which can be visualised in more
detail by reference to Fig. 5.

Fig. 6 shows how the CG of the downwelling light at
O.6 m and the upwelling light at O m varied over the year,
the CGs in each case being calculated from readings expressed as a proportion of the surface light. Fig. 6 also
shows the surface water temperature at this site, and various meteorological data for Gatwick Airport, which is 10
miles from the lake, obtained from the Meteorological
Office.

It can be seen that on the whole the CG of the down-

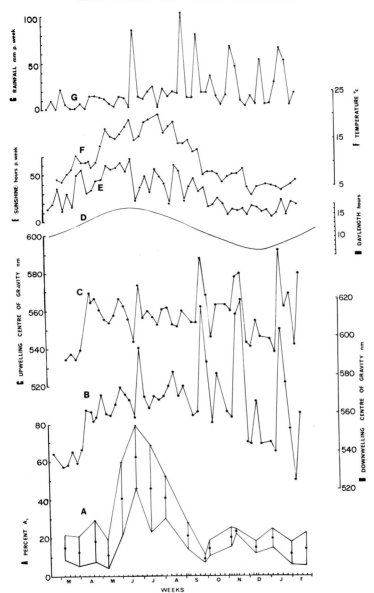

Fig. 6. Seasonal variations in the visual pigments and various environmental factors. A: mean percent A_1-based pigment and 95% limits for the mean. B: centre of gravity of downwelling irradiance at 0.6 m. C: centre of gravity of upwelling irradiance at 0 m. D, E, and G: various meteorological data for Gatwick Airport. F: surface water temperature.

welling light moved to longer wavelengths during the summer. There was, however, considerable variability: in particular, the results show a number of marked peaks during
the latter part of the year. These peaks were due to suspended mud, and were associated with the periods of heavy
rain that occurred at this time. The CG of the upwelling

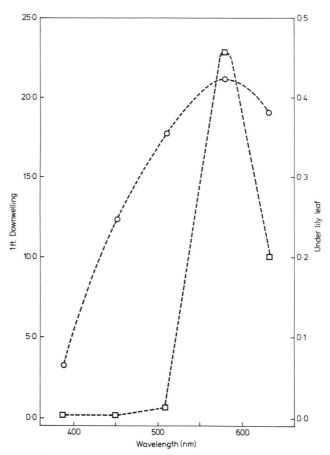

Fig. 7. Downwelling irradiance at 0.3 m in open water (circles), and
immediately beneath a lily leaf (squares) on 17th May 1975. The ordinate values are in quanta sec^{-1} cm^{-2} nm^{-1}.

light paralleled that of the downwelling light, though the
extent of the variations was smaller. This is to be expected since the upwelling light was measured at 0 m, whereas

the downwelling light was measured at a depth of 0.6 m,
so that the average amount of water that the light has
passed through is smaller in the former case than the
latter.

   Spectral measurements made in different parts of the
lake usually agreed well. For example, on 25 different
days when measurements were made at two fixed sites the
0.6 m downwelling CGs correlated highly together (r =
0.859). On a few occasions, however, large differences
occurred between different parts of the lake. These were
usually associated with periods of rain bringing down mud,
which often resulted in parts of the lake becoming heavily
discoloured, while other parts remained relatively clear.
The fish can also of course alter the spectral composition
of the light in their immediate vicinity by their own movements. In warm weather, for example, they were frequently
to be found lying just beneath the lily-pads, which obviously has a large effect on the light (Fig. 7). Marked
changes are also known to occur in the spectral composition
of the light reaching the surface of the water at twilight,
which result in a brief increase in the ratio of short
wavelength to long wavelength light at this time (Johnson
et al., 1967; Seliger and Fastie, 1968; Munz and McFarland
1973). This effect, which may be particularly significant
for any consideration of the function of the scotopic visual pigments, could be readily detected at Doctor's Lake.
On 14th March 1974 for example, the CG of the downwelling
light reaching the surface of the lake at 1600 h was 513
nm, and it remained between 513 nm and 518 nm until 1725 h.

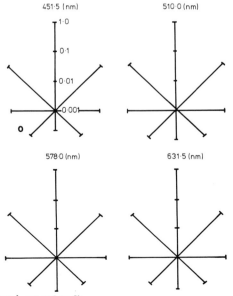

Fig. 8a (see legend overleaf)

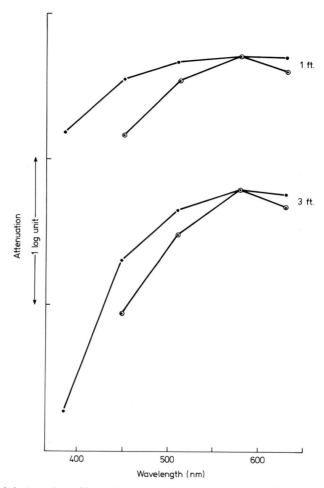

Fig. 8b (*a*) Angular distribution of radiance at different wavelengths on 5th September 1975. In each diagram the radiance is expressed as a proportion of the radiance at the zenith, and plotted on log axes. The central point on each diagram represents 0.001. (*b*) Downwelling (filled circles) and upwelling (open circles) irradiances compared for 0.3 m and 0.9 m data obtained on 29th September 1974. The vertical positions of the curves have in each case been adjusted to coincide at 578 nm in order to facilitate the comparison.

At 1730 h the CG had fallen to 499 nm, at 1800 h to 490 nm, and by 1815 h to 461 nm, after which there was insufficient light to make further measurements. Official sunset on 14th March 1974 was at 1804 h, but will have been earlier at the lake since this lies in a hollow.

Radiance Measurements Radiance measurements were made, with the opal Perspex cosine collector removed, at 45° steps round the vertical plane. Sample polar diagrams are shown in Fig. 8a. The form of the polar diagrams varied with wavelength, with intensity falling off faster as the readings were made further from the zenith at wavelengths both shorter and (to a lesser extent) longer than 571 nm. The upwelling light was therefore more monochromatic than the downwelling light. This could also be seen in the irradiance measurements (e.g. Fig. 8b).

Absolute Light Levels Most of the data reported so far refer to underwater readings relative to the surface illumination, and therefore reflect the characteristics of the water itself. The absolute light levels encountered in the lake over the year will however depend on other factors as well, such as the degree of cloud cover and the elevation of the sun.

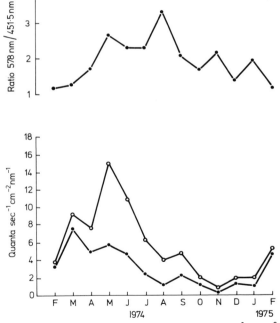

Fig. 9. Lower half: Mean downwelling quanta sec^{-1} cm^{-2} nm^{-1} at 451.5 nm (filled circles) and 578 nm (open circles) for each month, measured at 1200 h GMT at 0.6 m. Upper half: Ratio the number of quanta at 578 nm to the number at 451.5 nm, constructed from the data shown in the lower half.

Fig. 9b shows, for each month, the average number of quanta sec^{-1} cm^{-2} nm^{-1} at two wavelengths, for the downwelling irradiance at 0.6 m at 1200 h GMT, averaged over the different measurements made during that month. It can be seen that the highest light levels occurred in May, which was earlier than the peak daylengths and temperatures of the year. Fig. 9a shows the ratio of the absolute amounts of light at 0.6 m at 578 nm and 451.5 nm. Comparison of Figs. 6, 9a, and 9b shows that the absolute light intensity at 0.6 m also peaked earlier than the changes in the spectral quality of the light, whether these be measured relative to the surface light or in absolute units.

Pigment Extractions

Seasonal Variations   Fig. 6A shows the mean percentage of A_1-based pigment for the different samples of fish, together with the standard deviations of the means. Further details are given in Table 2. In agreement with expectations, there was an increase in the percentage of the A_1-based pigments during the summer months.

TABLE 2

Summary of Visual Pigment Extract Data

Date of collection	Mean Percent VP507$_1$	S.D. of the mean	Meanλ max.	Number of fish in sample
8.3.74	15.1	6.8	531.9	13
29.3.74	12.8	8.2	532.0	19
22.4.74	18.4	11.2	530.8	12
10.5.74	11.4	7.3	533.1	12
31.5.74	40.6	19.0	523.0	10
21.6.74	62.2	17.0	517.5	8
12.7.74	45.7	23.0	520.6	14
2.8.74	40.8	11.0	523.5	7
1.9.74	20.7	7.0	528.3	13
4.10.74	8.8	2.0	534.0	5
11.10.74	14.1	4.0	532.0	4
12.11.74	20.0	5.0	530.3	5
17.11.74	23.1	1.2	530.6	6
10.12.74	14.5	3.0	531.6	12
10.1.74	19.3	5.0	530.1	12
31.1.75	11.5	6.0	533.1	15
21.2.75	14.1	9.0	532.8	10

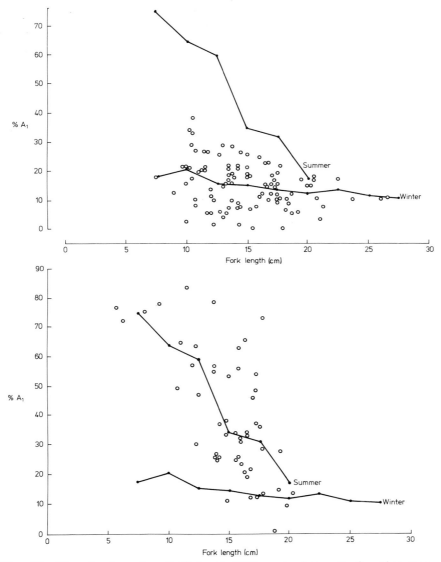

Fig. 10. Visual pigments of fish of different sizes. Each point represents an individual fish. The winter (March 8 - March 10; November 4 - January 1) fish are shown separately from the summer (May 31 - September 1) fish for clarity. The continuous lines show the means for each size class (2.5 cm etc.).

Individual Variations   There was a considerable variation in the composition of extracts prepared from different

individuals caught at the same time, which was partly related to the fact that the samples contained fish of different sizes. This can be seen from Fig. 10a and 10b, which
plot the percentage of the A_1-based pigment against fork
length for the individual fish. The summer and winter fish
are plotted on separate figures for clarity. It can be
seen that as the size (and presumably age) of the fish
increased both the amount of A_1-based pigment at a given
time of year, and the extent of the seasonal variation in
the two pigment types, decreased. These results confirm
in detail the findings of Bridges and Yoshikami (1970).

Stomach Contents   The stomach contents were noted for a
sample of 38 fish, caught on 13th June 1975. This analysis
was made because there is evidence that some other cyprinids change their diets as they get older, feeding less at
the surface on aerial insects, and more on vegetable matter near the bottom (see Discussion). No attempt was made
to carry out the analysis in any detail, the contents being
simply classified in each case on whether they contained
substantial amounts of algae, and on whether they contain
appreciable (over two) numbers of flying insects (easily
recognised by their wings). Two of the stomachs contained
exclusively duckweed, which also indicates surface feeding. The results are shown in Table 3 for the different
size classes of fish. Algae were more common in large fish,
and insects more common in small ones.

TABLE 3

*Number of animals in each size range having substantial
quantities of algae, flying insects, or duckweed
in their stomachs. The figures in brackets give the numbers
as proportions, excluding those fish that had empty stomachs.*

Fork Length cm	N	algae	insects	duckweed	nothing
Over 17.8	6	4(.80)	2(.40)	-	1
15.2-17.8	7	5(.72)	4(.57)	-	-
12.7-15.2	10	1(.13)	7(.88)	1(.13)	2
10.2-12.7	4	1(.25)	3(.75)	-	-
7.6-10.2	5	1(.25)	3(.75)	-	1
5.1-7.6	6	2(.50)	3(.75)	1(.25)	1

Although the sample was small,,and represents only one
day out of the year, the results suggest that the rudd
changes its feeding habits with age in the same way as
some other cyprinids.

Discussion

Seasonal Variations in the Light Climate in Freshwater

There have been comparatively few studies on seasonal variations in the spectral composition of the light in freshwater lakes. However, in general a higher mean extinction, or lower transparency as measured with a Secchi disc, correlated with an increase in the relative amount of long-wavelength light present (e.g. Juday and Birge, 1933; Vollenweider, 1961; Talling, 1971; Fig. 5 of this paper). Such measurements, which have been made on a seasonal basis on several occasions (e.g. Pearsall and Ullyott, 1934; Tressler *et al.*, 1940; Talling, 1971), show that there is considerable variation between lakes, with, however, a tendency for the highest extinctions and lowest transparencies to occur in the summer, in association with the growth of phytoplankton. Juday and Birge (1933) measured the colour of the water in 17 Wisconsin lakes in both the summer and the winter using platinum-cobalt standards and found that 10 of the lakes were more coloured in the summer and 3 more coloured in the winter, while 4 showed no measureable change. Their studies also revealed a high correlation between transparency, colour, and the plankton content of the water.

A more detailed study on the seasonal variation in the spectral composition of the light was made by Vollenweider (1961), for Lake Maggiore in Italy. Light measurements were made at four wavelengths, selected by glass filters, over a 16 month period from September 1959 to January 1961. The results showed a clear tendency for the short (430 nm) and long (630 nm) wavelength vertical extinction coefficients, expressed as a proportion of the mean of the extinction coefficients at 430 nm, 530 nm and 630 nm, to vary in opposite directions, with the long wavelength being transmitted better relative to the short wavelength light during the summer. For the months from September to January measurements were made on two successive years, and these showed that there were marked differences in the light conditions in the two years. Blindloss (1976) has recently reported light measurements made over a four year period in Loch Leven. Considerable variation occurred between years, and it was again found that the extinction depended mainly on the phytoplankton crop density. Spectral measurements showed that the light attenuation was greatest at short wavelengths, and that maximum transmission usually occurred at about 590 nm.

The present data, which show an increase in the mean extinction and a relatively greater amount of long-wavelength light during the summer, therefore agree with the majority of the data previously reported for eutrophic freshwater lakes. It is likely, however, that oligotrophic lakes and rivers will be more turbid and contain relatively

more long-wavelength light during the winter, due to the
rainfall causing an increase in the amount of dissolved
material in the water.

Changes in the Visual System

The best known hypothesis on how the visual pigments
are adapted to the environment is the 'sensitivity hypothesis', which states that the visual pigments have evolved to match the spectral quality of the ambient light in
such a way as to maximize the number of quanta absorbed.
This hypothesis has received support in the case of deep-
sea animals, where, for a variety of groups, it has been
found that the maxima of the visual pigment absorption
curves are shifted towards short wavelengths as compared
to those of shallow-water forms, in agreement with expectations based on the spectral composition of light in deep
water (e.g. Dartnall, 1975 for a review). Attempts have
also been made to apply the sensitivity hypothesis to the
seasonal variations in the visual pigments of freshwater
paired-pigment species. It has been suggested, for example
that the increase in A_2-based pigments during the winter
may be an adaptation to reddening of the ambient light,
caused by the lower elevation of the sun (Schwanzara,
1967). Seasonal changes in the spectral transmissivity of
the water itself will, however, probably be more important
in controlling the spectral composition of the fish's ambient light.

Table 4 shows examples of the relative efficiency of
the two visual pigments of the rudd, covering the range
of conditions that were encountered during the year. Data
on the light conditions on the days chosen for these examples are given in Figs. 4, 5 and 8. To make the calcu-
lations the quanta sec^{-1} cm^{-2} nm^{-1} incident on the measuring apparatus at each filter's wavelength of maximum trans
mission were multiplied by the two pigments' relative ab-
sorptions at that wavelength. The resulting products were
plotted and freehand curves drawn through them: the area
under these curves then gives a relative measure of the
quanta absorbed by the pigments. The results for the A_2-
based pigment were then multiplied by 0.7 to correct for
its lower photosensitivity (Dartnall, 1968), and finally
the ratio of the A_2-based pigment's absorption to that of
the A_1-based pigment calculated, both before and after the
correction for photosensitivity. In view of the fact that
the light measurements were only made at five wavelengths,
the figures given in Table 4 are only approximate. Nevertheless, the table illustrates certain clear trends. In
the first place, the A_2-based pigment becomes relatively
more effective as the water becomes more coloured, and
also as the depth increases. Secondly, the A_2-based pigment is usually relatively more effective for upwelling
as opposed to downwelling light, although this advantage

may not hold for shallow water heavily discoloured with suspended mud (e.g. 15th November 1974). Finally, although the A_2-based pigment has its absorption spectrum displaced towards long wavelengths in comparison to the A_1-based pigment, and more appropriately located with respect to the spectral distribution of the light in the lake, this advantage is largely cancelled by its lower photosensitivity.

TABLE 4

Examples of calculated ratios of quanta absorbed by $VP535_2$ and $VP507_1$, before and after correction for the lower photosensitivity of A_2-based pigments. Ratios less than one show a greater efficiency of the A_1-based pigment, more than one a greater efficiency of the A_2-based pigment.

Date	Measurement	uncorrected A_2/A_1 ratio	corrected A_2/A_1 ratio
7th February, 1975	0.6 m downwelling	1.22	0.86
	zero m upwelling	1.25	0.88
3rd May, 1974	0.6 m downwelling	1.49	1.03
	zero m upwelling	1.58	1.11
15th November, 1974	0.6 m downwelling	2.80	1.96
	zero m upwelling	1.56	1.10
20th September, 1974	0.3 m downwelling	1.14	0.80
	0.3 m downwelling	1.23	0.86
	0.9 m downwelling	1.49	1.04
	0.9 m upwelling	1.57	1.10

The sensitivity hypothesis clearly cannot account for the seasonal changes that occurred in the visual pigments in terms of seasonal changes in the spectral quality of the light. With the exception of the short periods following rain, during which the lake was highly turbid with mud, the water transmitted relatively more long wavelength energy during the summer than during the winter, whereas the absorption spectrum of the visual pigments moved towards short wavelengths during this time. Similarly it is known that the dorsal part of the rudd's retina contains relatively more of the A_1-based pigment than the ventral part, and so absorbs light of shorter wavelengths (Muntz and Northmore, 1971; Denton *et al.*, 1971), whereas the upwelling light (which will reach the dorsal retina) contained relatively less short wavelength energy than the downwelling light (e.g. Fig. 8, Table 4). The times of year and retinal locations at which the A_1-based pigment occurs are therefore the opposite to those that would be predicted by the sensitivity hypothesis.

Although light measurements made at a given point in the lake failed to reveal changes that could be related by the sensitivity hypothesis to the changes in the visual pigments, it is nevertheless possible that the fish are exposed to appropriate light environments at different times of the year through their own movements. As the examples in Table 4 illustrate, A_1-based pigments are more efficient than A_2-based pigments in shallow water, even when the spectral characteristics of the water are such that the A_2-based pigment is more effective at greater depths. The 'correct' mixture of pigments in a given body of water thus depends not only on the spectral characteristics of the water, but also on the depth at which the fishes live. In agreement with this, it has been suggested on several occasions that in freshwater the presence of both A_1- and A_2-based pigments is characteristic of surface living fishes, whereas fish that live near the bottom tend to have A_2-based pigments alone (Schwanzara, 1967); Muntz and Northmore, 1971; Muntz, 1973).

The present data might therefore be compatible with the sensitivity hypothesis if small rudd tend to live nearer the surface than large rudd, and if the fish tend to live nearer the surface in the summer than the winter. In this context it is therefore of interest that both chubb (*Squalius cephalus*) and dace (*Leuciscus leuciscus*), two other closely related cyprinids, have been reported to feed more on aerial insects when young than when old, and during the summer than during the winter (Hellawell, 1971, 1974). As we have seen (Table 3), it is likely that rudd similarly feed more on aerial insects when they are young, and they must feed on them more in the summer than the winter since they are only available at the former time. Increased feeding on aerial insects therefore tends to correlate with those times at which the animals have a greater amount of A_1-based pigment, and if this type of feeding is a valid indication of life near the surface it is compatible with the suggested explanation of the distribution of the two pigment types.

An explanation in these terms is however only partly successful. In particular, it fails to account for the differential distribution of the two pigment types over the retina. It is also apparent from Table 4 that neither pigment type has a very great advantage over the other under any of the conditions that occurred during the year, and other mechanisms could readily lead to greater improvements in sensitivity. The optical density of the rudd's visual pigment, for example, varies between about 0.65 in the dorsal retina and 0.2 in the ventral retina (Denton, *et al.*, 1971), and such density variations will have a greater effect on the quantal catch than any alteration

in the proportion of the two pigment types. If the main
function of the seasonal alterations in the pigments is
to increase sensitivity it is difficult to see why the
optical density of the pigment is not greater as well.

The best known alternative to the sensitivity hypothesis
is Lythgoe's (1968) contrast sensitivity hypothesis, which
states that pigments having absorption curves that fail
to match the spectral characteristics of the background
(offset pigments) will on occasion produce a greater contrast between the target and the background than those
that do (matched pigments), thereby increasing the detec-
tability of the stimulus. The improvement will only occur
under appropriate circumstances, namely the detection, in
shallow water, of nearby objects that are brighter than
the background: in all other cases matching pigments will
be more efficient. In the case of the rudd the A_1-based
pigment may be considered as an offset pigment, and the
A_2-based pigment as a matched pigment. The A_1-based pigment occurs chiefly in the dorsal retina, and probably
only during those times that the fish are living near the
surface. This suggests that this pigment is specialized
for looking downwards through the water body, in which
case the contrast sensitivity hypothesis would predict
that nearby stimuli brighter than the background are particularly important when they are beneath the animals.

Although this has never been shown it may be true,
especially since the upwelling light is considerably less
intense than the downwelling light, so that objects beneath
the animal are more likely to be brighter than the background than objects beside or above the animal. Nevertheless, in shallow bodies of water such as those inhabited
by rudd, objects will frequently be seen against the bottom itself and will presumably often be darker than it,
in which case matched pigments would be more efficient.
While therefore the present data do not rigorously exclude
the contrast sensitivity hypothesis, other possibilities
should also be considered. It may even be that the occurrence of $VP507_1$ is not related to the spectral position
of its absorbance curve at all, but to its greater stability and photosensitivity as compared to the A_2-based pigment. Another possibility is that the changes in the extractable pigments are in themselves unimportant, but reflect functional changes that are occurring in the cone
pigments. Pigment extracts usually only provide information
on the characteristics of the rods, which are responsible
for scotopic vision. Rudd, however, are said to be diurnal
in their behaviour (Siegmund, 1969; Roberts, 1957). Although an early behavioural experiment failed to find any
effect of daylength on the photopic spectral sensitivity
of the rudd (Muntz and Northmore, 1970), measurements
using the electroretinogram did reveal an effect of daylength under photopic conditions (Northmore and Muntz,

1970) and recent microspectrophotometric work has shown directly that the cones as well as the rods can change their pigments (Loew and Dartnall, 1976). The seasonal changes in the visual pigments may therefore be important through their effects, at present unknown, on photopic as opposed to scotopic vision.

Acknowledgements

We thank the Sussex River Authority and the Southern Water Authority Sussex Area Fishery Section and Biology Laboratories for their advice and their unstinting help in catching the fish, Christs Hospital for permission to work at Doctor's Lake, Brian Drury for electronic assistance, and the Technical Staff of the Laboratory of Experimental Psychology at Sussex University for building the light measuring instrument. Many people helped with the light measurements, and we would like to express our thanks for this. The work was supported by a grant from the Science Research Council.

References

Allen, D.M. (1971). *Vision Res.* **11**, 1077-112.
Allen, D.M., McFarland, W.M., Munz, F.W. and Poston, H.A. (1973). *Can. J. Zool.* **51**, 901-914.
Beatty, D.D. (1969). *Vision Res.* **9**, 1173-1184.
Beatty, D.D. (1975). *In: Vision in Fishes*, (M.A. Ali, ed.), Plenum Press, New York and London.
Bindloss, M.E. (1976). *Freshwater Biol.* **6**, 501-518.
Bridges, C.D.B. (1965). *Vision Res.* **5**, 239-252.
Bridges, C.D.B. and Yoshikami, S. (1970). *Vision Res.* **10**, 1315-1332.
Bridges, C.D.B. (1972). *In: Handbook of Sensory Physiology, Vol VII/I, Photochemistry of Vision*, (H.J.A. Dartnall, ed.), Springer-Verlag, Berlin, Heidelberg, New York.
Crescitelli, F. (1972). *In: Handbook of Sensory Physiology Vol. VII/I, Photochemistry of Vision*, (H.J.A. Dartnall, ed.), Springer-Verlag, Berlin, Heidelberg, New York.
Dartnall, H.J.A., Lander, M.R. and Munz, F.W. (1961). *In: Progress in Photobiology*, (B.C. Christensen and B. Buchman, eds.), Elsevier, Amsterdam.
Dartnall, H.J.A. (1968). *Vision Res.* **8**, 339-358.
Dartnall, H.J.A. (1975). *In: Vision in Fishes*, (M.A. Ali, ed.), Plenum Press, New York and London.
Denton, E.J., Munz, W.R.A. and Northmore, D.P.N. (1971). *J. mar. biol. Ass. U.K.*, **51**, 905-915.
Hellawell, J.M. (1971). *Freshwat. Biol.* **1**, 369-387.
Hellawell, J.M. (1974). *Freshwat. Biol.* **4**, 577-604.
Johnson, T.B., Salisbury, F.B. and Connor, G.I. (1967). *Science*, **155**, 1663-1665.
Juday, C. and Birge, E.A. (1933). *Trans. Wis. Acad. Sci. Arts Lett.* **28**, 205-289.
Lythgoe, J.N. (1968). *Vision Res.* **8**, 997-1012.
Loew, E.R. and Dartnall, H.J.A. (1976). *Vision Res.* **16**, 891-896.

Moon, P. (1961). *The scientific basis of illuminating engineering.*
    Dover Publications Inc., New York.
Muntz, W.R.A. and Northmore, D.P.M. (1970). *Vision Res.* **10**, 281-291.
Muntz, W.R.A. and Northmore, D.P.M. (1971). *Vision Res.* **11**, 551-561.
Muntz, W.R.A. (1973). *Vision Res.* **13**, 2235-2254.
Muntz, W.R.A. and Northmore, D.P.M. (1973). *Vision Res.* **13**, 245-252.
Munz, F.W. and McFarland, W.N. (1973). *Vision Res.* **13**, 1829-1874.
Northmore, D.P.M. and Muntz, W.R.A. (1970). *Vision Res.* **10**, 799-816.
Pearsall, W.H. and Ullyott, P. (1934). *J. exp. Biol.* **11**, 89-93.
Roberts, J.G. (1957). Rudd: How to Catch Them. Herbert, London.
Schwanzara, S.A. (1967). *Vision Res.* **7**, 121-148.
Seliger, H.H. and Fastie, W.G. (1968). *J. mar. Res.* **26**, 273-280.
Siegmund, R. (1969). *Biol. Zbl.* **88**, 295-312.
Talling, J.F. (1971). *Mitt. Internat. Verein, Limnol.* **19**, 214-243.
Tressler, W.L., Wagner, L.G. and Bere, R. (1940). *Trans. Amer.*
    Microsc. Soc. **59**, 12-30.
Tyler, J.E. and Smith, R.C. (1967). *J. opt. Soc. Amer.* **57**, 595-601.
Tyler, J.E. and Smith, R.C. (1970). *Measurements of spectral irradi-*
    ance underwater. Gordon and Breach, New York.
Vollenweider, R.A. (1961). *Mem. Ist. Ital, Idrobiol.* **13**, 87-113.
Wyszecki, G. and Stiles, W.S. (1967). *Color Science: concepts and*
    methods, quantitative data and formulas. Wiley, New York.

COMPARATIVE ASPECTS OF THE ACTIVITY RHYTHMS OF TAUTOG, *TAUTOGA ONITIS*, BLUEFISH, *POMATOMUS SALTATRIX*, AND ATLANTIC MACKEREL, *SCOMBER SCOMBRUS*, AS RELATED TO THEIR LIFE HABITS

BORI L. OLLA AND ANNE L. STUDHOLME

U.S. Department of Commerce,
National Oceanic and Atmospheric Administration,
National Marine Fisheries Service,
Northeast Fisheries Center,
Sandy Hook Laboratory,
Highlands, New Jersey 07732 USA.

Introduction

Over the past 10 to 15 years, there has been an increasing proliferation of studies on rhythms of activity in both terrestrial and aquatic animals (for recent reviews on all aspects of rhythmicity see Harker, 1964; Aschoff, 1965; Bünning, 1967; Menaker, 1971; Mills, 1973; Palmer, 1976). The bulk of studies has demonstrated the persistence of rhythmicity under constant laboratory conditions, indicating the presence of endogenous control. However, in nature, rhythms are both synchronized and phased by environmental cycles. Consequently, while results from investigations dealing with internal control of rhythmicity may, at times, be directly applicable to the natural situation (see Enright, 1975 for discussion) from an ecological point of view, it is important that rhythms be understood within the context of the animal's natural habits. Identifying how rhythms are integrated into the life habits of an animal is often neglected, but is of extreme value in understanding how it lives in harmony with the environment.

Regarding the marine environment, there is, at present, a dearth of studies on rhythms (DeCoursey, 1976). This stems from a variety of causes, including the difficulty encountered in capturing and maintaining delicate forms within the laboratory in a healthy condition over prolonged periods. Also, interpretation of results is often hindered because knowledge of the natural habits and environmental requirements of many marine species is sparse.

In this present work, we have investigated, under laboratory conditions, the influence of light and temperature on rhythms of activity in three species of marine fish, tautog (*Tautoga onitis*), bluefish (*Pomatomus saltatrix*), and Atlantic mackerel (*Scomber scombrus*), which have been shown to possess activity rhythms which are correlated with the daily light cycle (Olla and Studholme, 1972; 1975a; Olla, Bejda and Martin, 1974; Olla *et al.*, 1975).

All three species are migratory, undergoing vernal and autumnal movements along the eastern coast of North America. The seasonal movements and general distribution of tautog (Cooper, 1966; Olla *et al.*, 1974) are much more restricted than those of either bluefish (Lund and Maltezos, 1970; Walford, unpublished) or Atlantic mackerel (Sette, 1950).

The tautog, a demersal species, is found in association with bottom relief which provides cover and is, therefore, limited to areas which have appropriate shelter sites. In contrast, bluefish and mackerel, although quite different taxonomically, are both pelagic species. While seasonal movements in these species are correlated with changes in photoperiod, temperature also plays an important role in their movements and distribution, although the precise way in which temperature exerts an influence is not always clearly understood.

We have attempted in this work to relate the findings on rhythmicity to the different strategies possessed by each species for dealing with changes in the environment. Results from both published studies, as well as those not previously reported, have been included.

Materials and Methods

Laboratory studies on adult bluefish, Atlantic mackerel and tautog were conducted under controlled conditions of light and temperature in a multi-windowed 121,000 litre seawater aquarium (Olla, Marchioni and Katz, 1967). Normally operated as a semi-closed system, water quality was maintained by continual filtration through sand, gravel and oyster shell. Modifications in the original design of the system have included the addition of a piping network along the bottom of the aquarium which was covered by an 0.6 m layer of sand and gravel. The sand bottom served as a natural substrate for the tautog and provided additional filtration for maintaining seawater quality.

Water temperature was controlled primarily by room temperature for studies on adult bluefish and for one elevated temperature experiment on Atlantic mackerel, and for one on tautog. A thermal exchanger with associated cooling and heating units providing temperature control from 1^0 to 35^0C, was installed and utilized for a second elevated temperature experiment on adult tautog and for low temperature experiments on Atlantic mackerel and tautog.

Diurnal changes in light intensity from morning through evening civil twilight were simulated by eight rows of fluorescent lights mounted on the side walls above the aquarium (Olla *et al.*, 1967). The first row was controlled

by an automated dimming ballast and motor-driven potentio-
meter, ensuring smooth onset of light and avoiding abrupt
light increases which could cause startle responses in the
fish. Each row was turned on sequentially until reaching
peak intensity (3.5 x 10^3 mc for bluefish and mackerel
studies; 2 x 10^3 mc for tautog studies). The process then
reversed toward evening. Night-time illumination was provided by incandescent bulbs covered by diffusing screens
with light reflected off the ceiling. Light levels, as
measured at the water surface, averaged 2 x 10^{-1} mc.

For the study on the effects of seasonal changes in
temperature and photoperiod on adult tautog, eight fish
(three females, 48 to 58 cm; five males, 47 to 51 cm) were
introduced in August 1976 into the aquarium, with one male
(55 cm) from a previous study already in residence. Water
temperature and photoperiod in the aquarium were adjusted
to match the natural seasonal changes at Fire Island (Long
Island) New York where these fish were captured. From late
August until the first week in October, holding temperatures averaged 19.6°C (range 20.4° to 18.0°C) while photoperiod decreased from 13.3 to 12.5 h. Over the next 87
days temperature decreased (approximately 0.20°C/day)
until reaching 1.9° to 2.0°C at the end of December (10.0 h
photoperiod). Temperature averaged 2.1°C for 80 days until
the end of March with photoperiod increasing to 13.0 h.
Over the next 90 days, temperature was raised (approximately 0.15°C/day to 15.8°C while photoperiod increased
to 15.7 h.

During the same period, young tautog (12 to 15 cm)
were held in the aquarium in two rectangular PVC-framed
nylon mesh net cages (1.0 x 1.0 x 0.7 m; 10 fish/trap)
which rested on the sand bottom. Sufficient quantities
of *Mytilus edulis* were maintained for both adults and
young fish to allow *ad libitum* feeding.

Beginning 41 days after the fish were introduced, observations were made during daytime (0800 to 1500) and
periodically at night (2000 to 0300) for 10 min. each
hour, recording for all fish: 1) number of ingestions;
2) number of aggressions; 3) number of min. each fish
spent moving about or swimming expressed as percent of
time active for all fish. Observations were made in 4-day
periods with intervals of 3 to 4 days during which no
measurements were made.

For the experiment on the effects of elevated temperature on adult tautog two tests were conducted on two groups
of fish each consisting of two males and one female (47
to 55 cm). Each group was acclimated at 18.8° and 20.8°C
for 50 days and 80 days, respectively, under a constant
photoperiod of approximately 12 h prior to the increase
in temperature. In both tests temperature was raised
(0.04° C/h) and held at 28.7°C for 11 days, and then decreased (0.04 to 0.05°C/h) to acclimation levels. Obser-

vations of activity were made for 15 min. each hour for
12 h (0700 to 1800) in 4-day periods (Olla *et al.*, MS).
    Measurements of activity and rhythmicity of juvenile
bluefish were made in a 1500 litre fiberglass aquarium
as described in Olla and Studholme (1975a). Five juvenile
bluefish (16.5 to 19.2 cm) were captured by seine nets in
Sandy Hook Bay and held for 40 days at 20.4 ± 0.3 C under
a constant 12-h photoperiod. Every 2 days, the fish were
fed small pieces of clam until satiated.
    For observations of activity five successive stopwatch
readings were made each hour of the time for the lead
fish of the group to swim a measured distance (61 cm)
with the median (expressed in cm/s) used as the hourly
reading.
    For materials and methods for studies on rhythmicity
in bluefish see Olla and Studholme (1972); for bluefish
temperature studies see Olla and Studholme (1971); for
temperature studies on Atlantic mackerel see Olla *et al.*,
(1975); for field procedures on adult tautog, see Olla,
Bejda and Martin (1974).

Results and Discussion

    Adult tautog, *Tautoga onitis*, observed in the laborat-
ory were active during the day swimming about the aquarium, resting intermittently on the bottom, feeding, and
engaging in social interactions involving aggression
(Olla *et al.*, MS). At night, they were inactive, remaining quiescent and generally unresponsive to altering
stimuli in a state which resembled mammalian sleep (Fig.
1). These activity patterns agreed with what had been
observed under natural conditions, either directly by
using SCUBA or remotely, by tracking fish to which an
ultrasonic tag had been affixed (Olla, Bejda and Martin,
1974). The pattern of activity by day and inactivity at
night is one which apparently prevails for all members
of the labrid family (Hobson 1965, 1968, 1972; Stark and
Davis, 1966; Tauber and Weitzman, 1969; Collette and
Talbot, 1972). The only exception to the fish being quiescent at night occurred under natural conditions during
June, the peak spawning month for the species. Sonically
tagged fish seemed to move about at night (Olla *et al.*,
1974) and all adults, when approached by divers, were
more responsive and much more difficult to capture with
a hand-held net than during July through October. One
possibility for this modified behaviour was thought in
some way to be related to the spawning condition of the
fish. However scant the evidence was, it nevertheless
indicated that under certain conditions, adult tautog
might possess the capability for night-time activity, an
unusual attribute for a labrid.
    In nature, adult tautog during their active period,
move out and away from shelter as far as 500 m and then

return to spend the night (Olla *et al.*, 1974). We hypoth-
esized that if adult tautog were subjected to an environmental perturbation such as an elevation in temperature,
they might possess the capability of avoidance such as
shown by the pelagic species (see below).

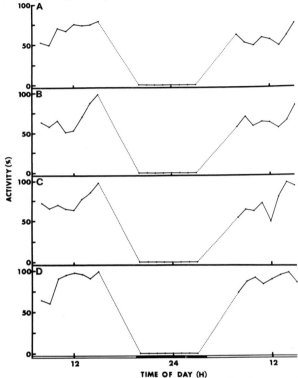

Fig. 1. Hourly activity of nine adult tautog, *Tautoga onitis*, measured
over a 4-week period under the following average temperatures and
photoperiods: (A) 13.9°C, 15.4 h; (B) 14.2°C, 15.5 h; (C) 14.5°C,
15.6 h; (D) 15.8°C, 15.7 h. Intervals between day (0800 to 1500) and
night (2000 to 0300) observations are indicated by a broken line
(----).

When adults were subjected to high temperatures
(28.7°C) the response, rather than reflecting avoidance,
was an overall decrease in activity (Fig. 2, Olla *et al.*,
MS). We surmised that the level and duration of the temperatures to which the fish were subjected were abnormal
relative to what they would experience during their evolutionary history. Under natural conditions the temperatures imposed would have been highly transient, lasting
but a few tidal cycles or at most several days. With shelter being such an essential requirement of the animal,

the strategy expressed of decreasing activity and thereby
remaining close to their homesite would be appropriate
under a natural perturbation, but obviously inappropriate
under a novel situation of prolonged exposure.

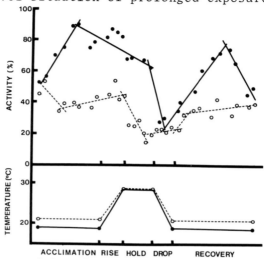

Fig. 2. Daytime activity of adult tautog, *Tautoga onitis*, in two separate high temperature experiments. Each point represents the mean
activity of three fish for 2 days. (Adapted from Olla *et al.*, MS).

In an earlier study (Olla and Studholme, 1975a) we had
interpreted the response of young tautog which remained
primarily within shelter under elevated temperature, as
a survival strategy.

We obtained very different results during the fall and
winter of 1976 when we subjected a group of nine adults
to decreasing temperature and photoperiod, which coincided
with the normal seasonal changes. Under natural conditions,
during late October and early November, adult tautog move
offshore to overwinter (Cooper, 1966; Olla *et al.*, 1974).

During this period, at 15° to 17°C, in 1971 and 1972
(Olla *et al.*, 1974) and 1974 (Olla, unpublished), divers
directly observed adult fish in the vicinity of their
homesite, where they were found throughout the summer.
Eleven to 23 days later, with temperature averaging 10°C,
adults were no longer observed. It is during this time
each year that commercial catches of tautog at the mouths
of estuaries show an increase (Olla, unpublished).

In the aquarium from October through early November
as the temperature decreased from 18° to 11°C and photoperiod decreased from 12.5 to 11.0 h, activity, feeding,
and aggression steadily decreased (Fig. 3). Then by the
second week of November when temperature reached 10.2°C,
there was an abrupt increase in swimming activity (Fig. 3).

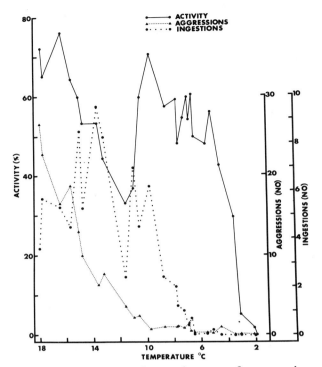

Fig. 3. Mean daytime activity (●——●), no. of aggressions (▲——▲) and number of ingestions of *Mytilus edulis* (●--●) for nine adult tautog, *Tautoga onitis*, under decreasing temperature (18° to 2°C) and photoperiod (12.5 to 10.0 h). Each point represents the mean of 2 days.

Associated with this increase, the fish no longer moved as individuals around the tank, but instead swam in groups of five to nine fish, generally remaining within 1 m of the surface. Feeding and aggression which had been diminishing, finally ceased (Fig. 3). The increased activity and schooling behaviour coincided with the normal offshore movement of adult fish.

To determine whether this rather gross change in behaviour might be reflected in some fashion during the night-time, beginning in early December, with temperature between 6° and 7°C, and photoperiod about 10.1 h, we began observing the fish at night. Rather than settle in, as they would normally do in the evening, the fish continued to swim about the aquarium in a school. In fact, for 48 h the rhythm was reversed, with more fish being active for longer periods at night than during the day (Fig. 4A). Over the next 2 weeks with temperature continuing to drop to 3.5°C, the fish still were active at night, but the overall level of activity for both day and night had decreased (Fig. 4B). Under natural conditions by the time temperature reached this level, the movement from inshore areas would have been completed.

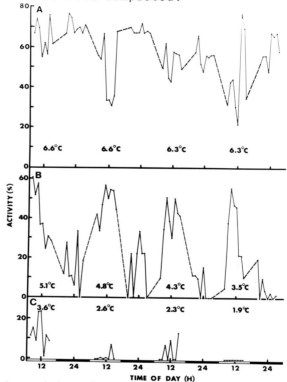

Fig. 4. Hourly activity of nine adult tautog, *Tautoga onitis*, under decreasing temperature and photoperiod: (A) 6.6^0 to 6.3^0C; 10.1 h; (B) 5.1^0 to 3.5^0C, 10.05 h; (C) 3.6^0 to 1.9^0C, 10.0 h. Intervals between day (0800 to 1500) and night (2000 to 0300) observations in (A) and (B) indicated by broken line (-----).

Although below 3.5^0C only intermittent observations were made at night, activity continued to decrease as temperature dropped with the fish finally burying under two fish traps that were in the aquarium and ceasing all activity at about 2^0C (Fig. 4C). They remained in torpor for 80 days. Then, by the beginning of April when the temperature had risen to 4^0C and photoperiod increased to 13.3 h, the fish began to emerge and become active. By the time temperature reached 9^0 to 10^0C (14.8-h photoperiod) all the fish had resumed a typical rhythmic pattern of daytime activity and night-time quiescence (Fig. 5).

Young tautog held in the aquarium during the same

period, in contrast with the adults, did not show any increase in activity when the temperature had dropped to
10.2°C. Then, as the temperature continued to drop they
decreased activity with some going into torpor at 4° to
5°C and the entire group by 2°C. The fact that they did
not increase activity as did the adults, would tend to
agree with their habits under natural conditions where
they have been shown to remain inshore to over-winter
(Olla *et al.*, 1974).

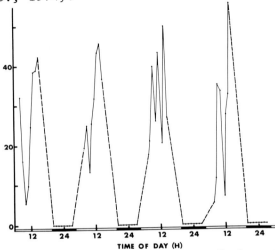

Fig. 5. Hourly activity of nine adult tautog, *Tautoga onitis,* measured
for 4 days under a 14.8-h photoperiod at 10.0°C. Intervals between
day (0800 to 1500) and night (2000 to 0300) observations are indicated
by a broken line (-----).

Even though these observations, made under decreasing
temperature and photoperiod, were on a single group of
adult fish, the comparison of laboratory results with
what is known about the natural life habits of these animals leads us to conclude that what we observed in the
aquarium was reflective of the natural situation. Strategies were expressed in response to these natural seasonal
changes which seemed to take precedence over the typical
day-active, night-inactive rhythm.
    Although the observed change in behaviour appeared to
be correlated with decreasing temperature and photoperiod,
it is possible that other factors, both internal and external, may have been involved.
    Transient changes in rhythmicity also occurred in the
bluefish (*Pomatomus saltatrix*) and the Atlantic mackerel
(*Scomber scombrus*) in response to temperature. These highly
migratory species, rather than being closely associated
with a particular locus like the tautog, are much more

influenced in their distribution by changing environmental
conditions, especially temperature (Olla and Studholme,
1975b; Olla *et al.*, 1975) and availability of food which
may vary markedly from day to day in particular locales.
   Thermal edges may impose barriers to the movements of
these fish (for Atlantic mackerel, see Sette, 1950). In
addition, fish may become entrapped within a thermal regime
which can exceed physiological limits for the species. In
the case of bluefish and Atlantic mackerel, they appear to
possess the capability of regulating body temperature behaviourally by avoidance (Olla and Studholme, 1971; 1975b;
Olla *et al.*, 1975). Once encountering and then sensing a
thermal edge which is potentially stressful, the fish must
simply swim away to avoid the stress. Activity increases
would be highly transient and the effect on the 24-h rhythm
minimal. Recent studies on juvenile bluefish have shown
just this kind of response (Olla, unpublished).
   This capability of behaviourally regulating body temperature by selecting water temperature (behavioural thermo-
regulation) has been shown for other species both *in situ*
and under controlled laboratory conditions (Rozin and
Mayer, 1961; Neill, Magnuson and Chipman, 1972; Neill and
Magnuson, 1974).
   However, in the case of entrapment, which implies at
least a local homeothermal environment, the fish will increase activity as they attempt avoidance. Our observations
on adult bluefish and Atlantic mackerel were made in this
type of homeothermal environment. From these observations
we wish to show that the priority to avoid will supercede
rhythmicity at least to the point where the animals are
still able to afford the additional energy needed during
the time when their activity would normally be reduced.
We will begin our discussion of bluefish and Atlantic
mackerel by describing their daily rhythms and then the
effect temperature has on these with respect to their
ability to thermoregulate.
   Bluefish and Atlantic mackerel, quite different taxonomically, swim continuously in a clearly defined rhythm,
swimming faster by day than by night (Fig. 6A, B, C, Olla
and Studholme, 1972; 1975b; Olla *et al.*, 1975.) In addition,
a series of experiments on bluefish showed that the rhythm
of activity had an endogenous component (Olla and Marchioni, 1968; Olla and Studholme, 1972).
   At acclimation temperatures, within the preferred range
for each species, differences in swimming speed between
day and night were larger for adult bluefish than for
Atlantic mackerel, resulting in the amplitude of the rhythm
being greater. Variability in swimming speed also differed,
being higher for bluefish than mackerel. For example, in
a typical 4-day period of activity for each species (Fig.
6A, B), minute-to-minute coefficient of variation for
bluefish averaged 12.7% during the day; 28.2% at night

(Olla and Studholme, 1972); for Atlantic mackerel, 3.9% during the day; 3.8% at night. The more consistent pattern of swimming speed in mackerel may be attributable to the fact that adult mackerel must swim above a certain velocity to maintain position in the water column since they lack a hydrostatic organ and to ram gill ventilate (Roberts, 1975). Bluefish, possessing a swim bladder could swim at much lower speeds. Oxygen requirements at these speeds are apparently met by active gill ventilation.

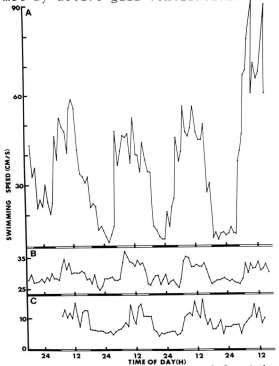

Fig. 6. Hourly median swimming speeds measured for 4 days for, (A) adult bluefish, *Pomatomus saltatrix*, under a 14.15-h photoperiod at 21.6°C; (B) Atlantic mackerel, *Scomber scombrus*, under a 10.42-h photoperiod at 12.8°C; (C) juvenile bluefish under a 12.0-h photoperiod at 20.4°C. (Adult bluefish data adapted from Olla and Studholme, 1972).

Schooling patterns in both species also followed a rhythmic pattern. For adult bluefish, the number of fish schooling was greater by day than by night (Fig. 7A, Olla and Studholme, 1972). For Atlantic mackerel, while group size was fairly consistent, the cohesiveness of the school, estimated by the average linear distance between fish (school gap) varied from day to night. In four of the six groups of Atlantic mackerel studied, the fish tended to

be grouped more tightly by night than during the day
(Fig. 7B) with the reverse pattern being characteristic
of the other two groups. We cannot as yet speculate on the
reasons for these differences.

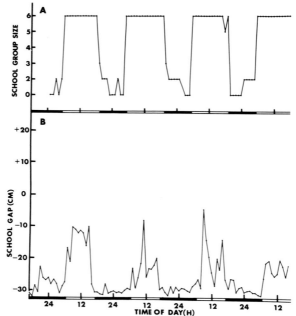

Fig. 7. Hourly medians of, (A) adult bluefish, *Pomatomus saltatrix*,
school group-size, and (B) Atlantic mackerel, *Scomber scombrus*,
school gap (average linear distance between individuals) measured
for 4 days under conditions as given in Fig. 6. (Bluefish data adapted
from Olla and Studholme, 1972).

   While the reduced activity at night might be interpreted
as a unique kind of rest period for fish continually in
motion, in contrast with the low responsiveness of tautog,
these animals were highly responsive to altering stimuli
at night. Surface disturbances, introduction of prey, or
the sudden onset of light were met with almost immediate
increases in activity, and depending on the stimulus introduced, the fish might begin feeding or show behaviour which
reflected escape. High levels of responsiveness and perception of changes in the environment were capabilities
these animals appeared to possess throughout the daily
rhythmic cycle.
   The effect of temperature on activity rhythms of bluefish and Atlantic mackerel was similarly manifested. In
both species when temperature rose or fell beyond certain
levels and activity increased, normal patterns of rhythmicity were affected.

In adult bluefish when temperature was gradually raised (mean 0.02°C/h) from an acclimation level of 19.9°C (average 16.0-h photoperiod) the fish showed avoidance by in-

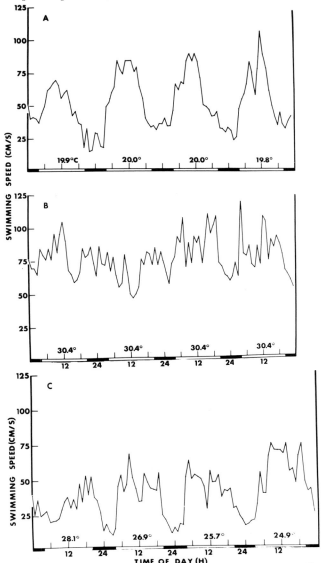

Fig. 8. Hourly median swimming speeds of adult bluefish, *Pomatomus saltatrix*, during high temperature test: (A) 19.8° to 20.0°C; (B) 30.4°C; and (C) 28.1° to 24.9°C. Photoperiod averaged 16.0 h. (Figure as presented in Olla and Studholme, 1971).

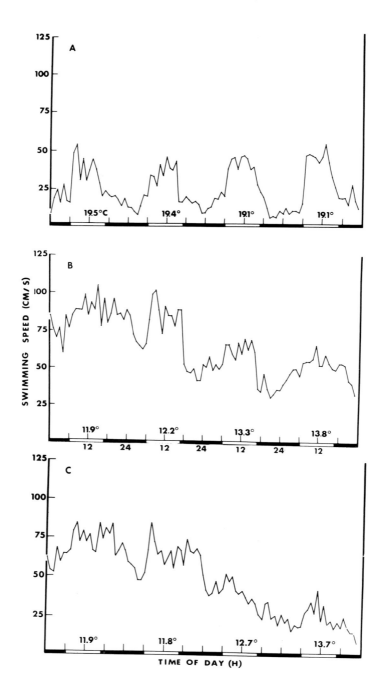

SWIMMING SPEED (CM/S)

TIME OF DAY (H)

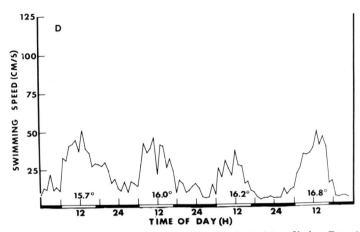

Fig. 9. Hourly median swimming speeds of adult bluefish, *Pomatomus saltatrix*, during low temperature test: (A) 19.1⁰ to 19.5⁰C; (B) 11.9⁰ to 13.8⁰C; (C) 11.8⁰ to 13.7⁰C; (D) 15.7⁰ to 16.8⁰C. Photoperiod averaged 10.7 h. (Figure as presented in Olla and Studholme, 1971).

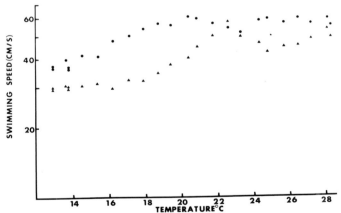

Fig. 10. Mean day (●) and night (▲) swimming speeds of Atlantic mackerel, *Scomber scombrus*, during high temperature test. Each point represents the mean of 3 days.

creasing activity (Olla and Studholme, 1971). At high temperature (30.4⁰C), in contrast with the typical pattern of rhythmicity evident during acclimation (Fig. 8A), rhythmicity was disrupted for 48 h, with the normal day-night cycle apparently shifting, i.e., the fish briefly (for 4 to 5 h) decreasing activity around midday and averaging higher speed that night than during the following day (Fig. 8B). For the remaining 48 h at this temperature, although activity remained high, the fish once again swam faster during the day than at night (Fig. 8B). As the temperature was lowered to 28.1°C activity decreased so sharply that the rhythm was again disrupted as the fish continued to swim at low speed for 7 h after light onset (Fig. 8C). Although for several days, as the temperature continued to drop average speeds remained lower than during acclimation, the rhythm was clearly evident (Fig. 8C).

When bluefish were subjected to decreasing temperature (mean 0.01°C/h) from an acclimation level of 19.5°C (average 10.7-h photoperiod; Fig. 9A), rhythmicity was also affected (Olla and Studholme, 1971). At low temperatures (11.8°; 11.9°C) although the fish continued to show the typical increase in swimming speed at light onset, the normal decrease in activity at the onset of the dark period did not occur. Instead, the fish continued to swim at day speeds for 8 to 9 h before decreasing activity (Fig. 9B, C). Typical rhythmic patterns returned after the temperature was raised above 12° to 13°C (Fig. 9B, C, D). At both high and low temperature extremes, during the periods when the rhythm was disrupted, the 'resting phase' (determined by a significantly prolonged decrease in swimming speed) lasted only about 4 to 6 h as the fish maintained higher speed for most of the daily cycle.

Rhythmicity was also disrupted in Atlantic mackerel at temperature extremes. When temperature was gradually increased (mean 0.02°C/h) from an acclimation level of 13.3°C (average 15.3-h photoperiod) mackerel first exhibited avoidance behaviour at 14° to 15°C, increasing speed during the day but not at night (Fig. 10; Olla *et al.*, 1975). The amplitude of the rhythm had increased. Then, as temperature rose above 18°C, and avoidance was obviously not possible within the homeothermal environment of the aquarium, the fish began to increase speed at night (Fig. 10). Between 21° and 24°C, in contrast with normal rhythmicity during acclimation (Fig. 11A), there were periods when the rhythm was reversed, as the fish swam faster by night than by day, or absent, with the fish swimming about the same speed day and night (Fig. 11B). As the temperature continued to rise (eventually to lethal levels with 4 of 8 fish dying by 28.6°C), the rhythm was re-established (Fig. 11C). We intepreted this as resulting from the fact that the fish were swimming at the maximum levels they could maintain over a lengthy period and could no longer afford the additional expenditures of energy needed to swim at night at the highest speed (Olla *et al.*, 1975).

At low temperature (2.7° to 1.9°C) following acclimation at 7.9°C (average 15.3-h photoperiod) similar shifts occurred (Fig. 12A, B, C). These temperatures were lethal for eight of 13 fish, with rhythmicity re-established for the five surviving fish when temperature was raised to 7.6° to 7.8°C (Fig. 12D).

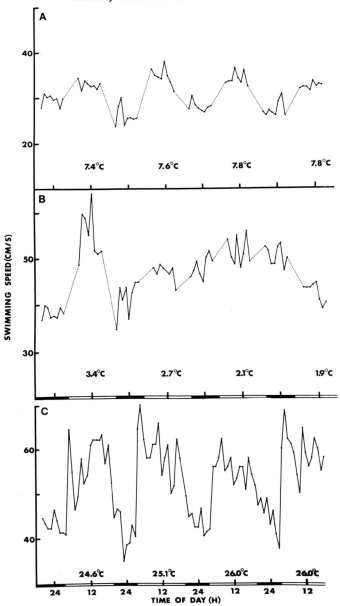

Fig. 11. Hourly median swimming speeds of Atlantic mackerel, *Scomber scombrus*, during high temperature test: (A) 13.4° to 14.6°C; (B) 22.5° to 23.6°C; and (C) 24.6° to 26.0°C. Photoperiod averaged 15.3 h. In-terval between observations indicated by broken line (----).

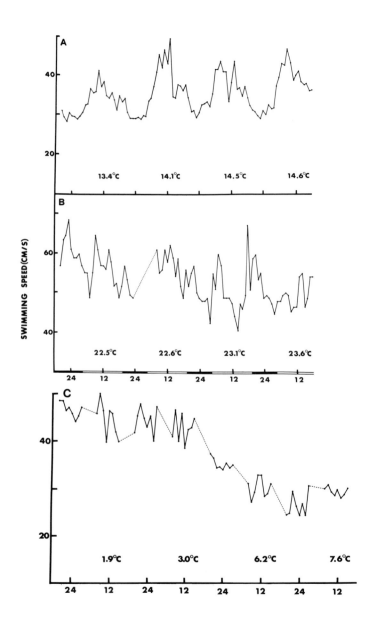

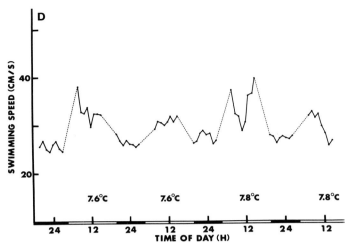

Fig. 12. Hourly median swimming speeds of Atlantic mackerel, *Scomber scombrus*, during low temperature test: (A) 7.4^0 to 7.8^0C; (B) 3.4^0 to 1.9^0C; (C) 1.9^0 to 7.6^0C; and (D) 7.6^0 to 7.8^0C. Photoperiod averaged 15.3 h. Intervals between day (0800 to 1500) and night (2000 to 0300) observations are indicated by broken line (----).

In summary, we have attempted to show how patterns of activity and the way in which these may be modified by environmental change relate to the ecological niche each species occupies. The pelagic bluefish and mackerel being more closely associated with particular thermal regimes rather than a specific place, respond to either high or low temperature extremes with avoidance. In contrast, when tautog were subjected to elevated temperature extremes, they did not show changes in behaviour reflective of avoidance. The high temperature to which the fish were subjected when occurring at all under natural conditions, would be highly transient. Thus remaining in proximity to shelter at temperatures which are not lethal over the short term, would seem more appropriate than avoidance in a species for which shelter is ecologically limiting. On the other hand when tautog were subjected to decreasing temperature and photoperiod but which closely followed natural seasonal changes, activity not only increased but extended into the normally quiescent night period. This change coincided with the natural seasonal movements of tautog off-shore. The response in this case was related to a shift in seasonal habitat requirements.

Acknowledgements

We wish to express our grateful appreciation to Allen J. Bejda, Carol Samet, and A. Dale Martin for their assistance throughout all phases of these studies.

150   B.L. OLLA ET AL.

    This work was supported, in part, by grants from the
U.S. Energy Research and Development Administration,
No. E (4907) 3045.

References

Aschoff, J. (1965). (Editor). *Circadian clocks*. North-Holland Publish-
    ing Co., Amsterdam, 479 pp.
Bünning, E. (1967). *The Physiological clock*. Springer Verlag, Berlin,
    167 pp.
Collette, B.B. and Talbot, F.H. (1972). *In: Results of the tektite
    program: ecology of coral fishes*, (B.B. Collette and S.A. Earle,
    eds.), *Los Angeles Co. Mus. Sci. Bull.*, **14**, 125-170.
Cooper, R.A. (1966). *Trans. Am. Fish Soc.*, **95**, 239-247.
DeCoursey, P.J. (1976). (Editor). *Biological rhythms in the marine
    environment*. University of S. Carolina Press, Columbia, U.S.A.,
    283 pp.
Enright, J.T. (1975). *In: Marine ecology* (O. Kinne, ed.), Wiley -
    Interscience, London, pp. 917-944.
Harker, J.E. (1964). *The physiology of diurnal rhythms*. Cambridge
    University Press, London, 114 pp.
Hobson, E.S. (1965). *Copeia*. 1965, 291-302.
Hobson, E.S. (1968). *Res. Rep. Fish Wildl. Serv. U.S.*, **73**, 1-92.
Hobson, E.S. (1972). *Fishery Bull.*, *U.S.*, 70, 715-740.
Lund, W.A., Jr. and Maltezos, G.C. (1970). *Trans. Am. Fish. Soc.* **99**,
    719-725.
Menaker, M. (1971). (Editor). *Biochronometry*. National Academy of
    Sciences, Washington, D.C., 662 pp.
Mills, J.N. (1973). (Editor). *Biological aspects of circadian rhythms*.
    Plenum Press, London and New York, 319 pp.
Neill, W.H. and Magnuson, J.J. (1974). *Trans. Am. Fish. Soc.*, **103**,
    663-710.
Neill, W.H., Magnuson, J.J. and Chipman, G. (1972). *Science, N.Y.*
    176, 1443-1445.
Olla, B.L., Bejda, A.J. and Martin, A.D. (1974). *Fishery Bull. U.S.*
    72, 27-35.
Olla, B.L. and Marchioni, W.W. (1968). *Biol. Bull. mar. biol. Lab.*,
    Woods Hole, **135**, 530-536.
Olla, B.L., Marchioni, W.W. and Katz, H.M. (1967). *Trans. Am. Fish.
    Soc.* **96**, 143-150.
Olla, B.L. and Studholme, A.L. (1971). *Biol. Bull. mar. biol. Lab.*,
    Woods Hole, **141**, 337-349.
Olla, B.L. and Studholme, A.L. (1972). *In: Behaviour of marine ani-
    mals: current perspectives in research*, (H.E. Winn and B.L. Olla,
    eds.), Plenum Press, New York, pp. 303-326.
Olla, B.L. and Studholme, A.L. (1975a). *In: Proc. 9th Europ. mar.
    biol. Symp.*, (H. Barnes, ed.), Aberdeen University Press, Aberdeen,
    pp. 75-93.
Olla, B.L. and Studholme, A.L. (1975b). *In: Second joint U.S./U.S.S.R.
    symposium on the comprehensive analysis of the environment*, Environ-
    mental Protection Agency, Washington, D.C., pp. 25-31.
Olla, B.L., Studholme, A.L., Bejda, A.J., Samet, C. and Martin, A.D.
    (1975). *In: Combined effects of radioactive, chemical and thermal*

releases to the environment, Int. atom. Energy Ag. (SM-197/4), Vienna, pp. 299-308.

Olla, B.L., Studholme, A.L., Bejda, A.J., Samet, C. and Martin, A.D. (Manuscript). The effect of temperature on social behaviour and activity of adult tautog, *Tautoga onitis,* under laboratory conditions.

Palmer, J.D. (1976). *An introduction to biological rhythms.* Academic Press, New York, 375 pp.

Roberts, J.L. (1975). *Biol. Bull. mar. biol. Lab., Woods Hole,* **148**, 85-105.

Rozin, P.N. and Mayer, J. (1961). *Science, N.Y.,* **134**, 942-943.

Sette, O.E. (1950). *Fishery Bull. Fish Wildl. Serv. U.S.* **51**, 1-251.

Stark, W.A., II and Davis, W.P. (1966). *Ichthyol. Aquarium J.* **38**, 313-356.

Tauber, E.S. and Weitzman, E.D. (1969). *Commun. Behav. Biol.* **3A**, 131-135.

BEHAVIOURAL AND PHYSIOLOGICAL RHYTHMS OF FISH IN THEIR NATURAL ENVIRONMENT, AS INDICATED BY ULTRASONIC TELEMETRY OF HEART RATE

I.G. PRIEDE

Department of Zoology,
University of Aberdeen, U.K.

Introduction

Heart rate telemetry from free-living fish in the natural environment is a new development in fisheries biology having only emerged in the last few years as a practicable field technique (Priede and Young, 1977). Measurement of locomotor activity has been widely used in laboratory studies of circadian rhythms of fish (Spencer, 1939; Muller and Schreiber, 1967) and movements can be monitored in the wild by means of sonic tracking transmitters (Trefethen, 1956; Stasko, 1975). Richardson and McCleave (1974) however point out that indicator processes other than locomotor activity may be better suited for the study of circadian rhythms and Kneis and Siegmund (1976) suggest that heart rate is a more sensitive measure of the individual state than some visible behaviour patterns. The purpose of this paper is to review the capabilities of the heart rate telemetry technique, present some results from brown trout (*Salmo trutta* L), and to discuss the relationship between heart rate and the fish's behaviour and physiology.

The Heart Rate Telemetry Technique

Frank (1968) described the design and use of an FM radio transmitter for telemetry of the electrocardiogram of rainbow trout (*Salmo gairdneri*) in the laboratory. The transmitter was small enough to be carried by the fish and the ECG was detected by hook electrodes. The following year Nomura and Ibaraki (1969) published some preliminary results from FM radio telemetry of the electrocardiogram of free-swimming rainbow trout in a pond. The transmitter was fitted in a buoy which was towed behind the fish by means of trailing ECG leads.

Small radio telemetry ECG transmitters have since become commercially available and Nomura *et al.* (1972) used one in back-pack form on rainbow trout and Sockeye salmon

(*Oncorhynchus nerka*) in a hatchery, Roberts *et al.* (1973) tested one in a laboratory on goldfish (*Carassius auratus*). Radio transmission is severely attenuated in all but highly non-conductive waters and typical ranges are only of the order of a few metres even less in sea-water. The development of ultrasonic transmitters with ranges of over 100m, potentially up to 1000m, has been a major step forward.

Two types of ultrasonic transmitters have been demonstrated so far. Kanwisher *et al.* (1974) used a continuous ultrasonic carrier signal (40-80kHz) modulated in frequency by the ECG exactly as in conventional FM radio telemetry. Their transmitter was 70mm long and was inserted into the stomach of cod (*Gadus morhua*) (Wardle and Kanwisher, 1974). ECG leads were drawn out through the gill apertures and electrodes inserted in the region of the pectoral and pelvic girdles. Kanwisher *et al.* (1974) also tested this transmitter on plaice (*Pleuronectes platessa*) to which it was attached externally, as well as Salmon (*Salmo salar*) and skipjack tuna (*Katsuwonus pelamis*). They mention a new ultra-miniature version (35mm by 7mm) used on Pacific mackerel (*Scomber japonicus*) but no details are given. The power output of their transmitters was low giving a range of up to 100m, sufficient for work in laboratory tanks.

Young and Wiewiorka (1975) made great economies in power requirements for heart rate transmission by using the ECG to trigger single short ultrasonic pulses corresponding to each heart beat. Their transmitter is very compact (35mm by 8mm) and was used in the field work on brown trout (*Salmo trutta*) which is discussed below. The transmitter was attached externally to the dorsum of the fish (Priede and Young 1977) and a single ECG lead was connected to an electrode inserted in the pectoral region. One of the attachment wires of the transmitter 'tag' was used as an ECG earth electrode. The ultrasonic signal was at a relatively high frequency (220kHz) which requires a smaller emitting transducer but the signal is more attenuated than at lower frequencies (Kanwisher *et al.* 1974). The study area, Airthrey Loch, Stirling, Scotland (56°09 N, 3°55 W) is a relatively small body of water where at least one of three fixed hydrophone stations could pick up the signals throughout the fish's normal daily activity. Even in larger water bodies, including the sea, fish movements are often restricted and territorial (Holliday *et al.*, 1974; Hawkins *et al.*, 1974; Kelso, 1976, Warden and Lorio, 1975) so that continuous monitoring should be practicable for many species especially using sonar buoys to relay telemetry data to a shore base (Fig. 1). Such buoys could easily be moved should the fish alter its centre of activity. Various forms of boat - and ship-mounted hydrophones can be used (Stasko and Polar, 1973) and continuous contact has been maintained with migratory fish in the open sea for tracking purposes (Tesch, 1972; Walker *et al.*, 1971, 1977).

Depending on the trade-off in compromising between transmitter size, range and life, transmitter lives from a few
days up to a month are possible.

The heart beat telemetry is technologically feasible; the
performance of early prototype equipment could be improved
or developed in particular directions and it remains for
the biologist to formulate the right questions to ask in
investigations of this type.

The only practical field telemetry work to date is the
study on Airthrey Loch (Priede and Young, 1977; Priede,
1977b). The results will be discussed below from the point
of view of circadian rhythms of fish.

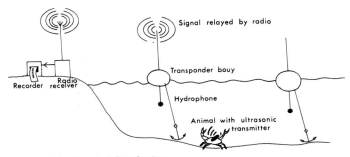

Proposed estuarine and sub-littoral system

Fig. 1. The main elements of a proposed physiological telemetry system for aquatic organisms. The range of the ultrasonic transmitter
attached to the animal is 100 to 400m. One or more sona buoys can be
positioned to relay the signal by radio to a convenient data logging
base.

Results

The heart rate telemetry data from four fish will be
considered in detail (Table 1). Fish A and B were monitored
at summer temperatures of about 15°C and the details of
their behaviour are described elsewhere (Priede and Young,
1977). C and D can be regarded as nominally winter fish
with water temperatures close to 5.5°C. Histograms of
heart rates throughout the period of monitoring for fish
C and D are shown in Figs. 2 and 3 respectively. The histograms are based on a one minute sample of heart beats taken
every 10 minutes so may omit some isolated events in the
record.

Fish C. This fish had a high heart rate during the first
24h but afterwards it settled into a diurnal pattern of
activity with high heart rates during the day and lower
at night (Fig. 4). For this analysis the demarcation between day and night is taken as civil twilight which is
defined as the time when the sun is 6° below the horizon.
This was shown previously (Priede and Young, 1977) to

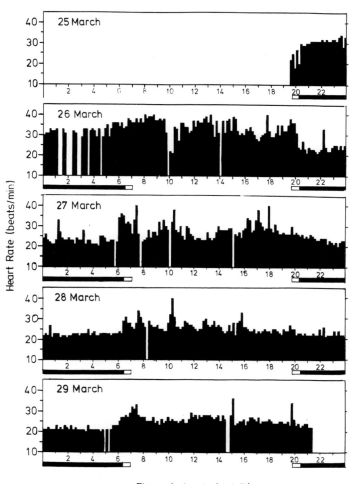

Time of day h (B.S.T.)

Fig. 2. Fish C. Histogram of heart rate taken at 10 minute intervals.
The time scale is British Summer time. The black horizontal bar below
each day's record indicates night time from civil twilight to civil
twilight. The ends of the unshaded extensions of the bar indicate
sunrise and sunset. Breaks in the histogram occur when the signal
from the fish was temporarily lost.

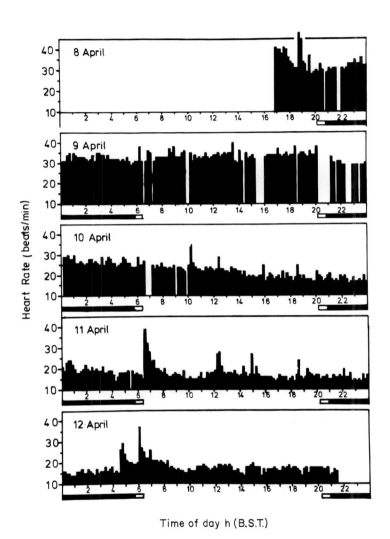

Time of day h (B.S.T.)

Fig. 3. Fish D. Histogram of heart rate as in Fig. 2.

give a better distinction between day and night heart
rates than either sunset/sunrise or nautical twilight.

TABLE 1

*Details of brown trout (Salmo trutta L) from which heart rate
telemetry was carried out. References are given to
papers in which aspects of this data is also discussed*

Fish	Weight	Length	Date of Release	Duration of Monitoring	Water Temp.	
A	514g	397mm	25 Jun 74	69.25h	15.5°C	(fish 3 in Priede and Young, 1977)
B	553	345	11 Sep 74	169.25	14.6	(fish 4 in Priede and Young, 1977)
C	366	310	25 Mar 75	96	5.8	(fish 2 in Priede, 1977b)
D	376	328	8 Apr 75	96	5.6	(fish 3 in Priede, 1977b)

The water temperature was uniform at 5.8°C throughout the
period of monitoring. The 24h period of recovery from tagging and release has come to be regarded as normal in fish
telemetry studies.

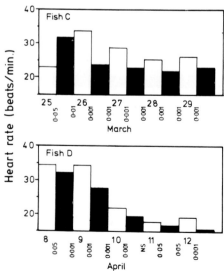

Fig. 4. Histograms showing the mean day and night heart rates for
fish C and D. The P values for level of significance of the difference
between adjacent pairs of means are given below each diagram (t test).
NS - Not significant.

The movement patterns of the fish after this recovery were of a normal 'home range' type with excursions from a centre of activity. (Young *et al.*, 1972; Holliday *et al.*, 1974).

Fish D This fish (Fig. 3) showed very high heart rates initially and spent all its time away from the presumed usual home range where it had been caught. It returned on the morning of the third day, adopted a home range and was very inactive for the rest of the tracking period. The comparison of mean day and night rates (Fig. 3b) does not show a distinct circadian rhythm although there are distinct dawn peaks in the continuous record (Fig. 3).

Data Analysis

Fig. 5 shows the heart rate data from fish A,B,C and D. The first 24h of each fish record has been omitted to exclude the anomalies associated with recovery from tagging and the mean heart rate for each hour of the day is shown in histogram form.

The mean time of day when heart-beats occur was calculated by the methods of Mardia (1972), and the uniformity of the data assessed by the Rayleigh test (Table 2).

TABLE 2

The mean time of day when a heartbeat occurred in the four Brown trout together with confidence limit parameters

Fish	Mean direction (time $\bar{x}_o$	Rayleigh Uniformity test $2n\bar{R}^2$	p	$\hat{k}$	95% Confidence limits on $\bar{x}_o$
A	10h 30min	10.68	.01	.0384	± 2h 25min
B	11h 25min	67.08	.001	.0582	± 54min
C	12h 19min	51.20	.001	.0971	± 1h 2 min
D	6h 25min	16.24	.001	.0793	± 1h 51min

The null hypothesis of a uniform distribution (no significant fluctuations in heart rate) is rejected for all 4 fishes. The mean time of heart beat occurrence in all cases anticipates solar noon (13.14 B.S.T.) although the data of fish D is obviously biassed by the dawn peak. A cardioid function (which approximates to the Von Mises distribution when k is small as in this case (Mardia 1972)) based on these estimates of the mode is fitted to the data in Fig. 5. By ranking chi-squared goodness-of-fit values of the data to the cardioid, uniform, and step functions of Fig. 5, the last is shown to be the most realistic, suggesting that it is not a 24h period sinusoid oscillator which directly determines heart rate. The fish responds to dawn and dusk transitions with greater cardiac

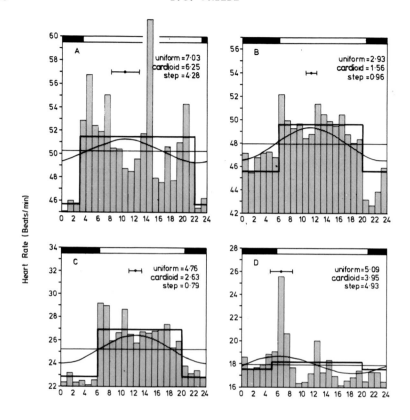

Fig. 5. Histograms of the mean heart rate during each hour of the day. Data from the whole period of monitoring for each fish is used excluding the first 24h. The horizontal black bar at the top of each diagram indicates night time from civil twilight to civil twilight. The dot and horizontal bar represent the mean time of occurrence of a heart beat $\bar{x}_O$ and its 95% confidence limits. Uniform, cardioid and step function distributions are shown fitted to each set of data and their respective chi-square values are given above each histogram.

activity by day than by night. The importance of the dawn transition stimulus is emphasised by the record for fish D.

The Circadian Rhythm of Heart Rate

The pattern of cardiac activity with high heart rates during the day corresponds in a general way to the patterns of locomotor activity discovered by means of sonic tracking. (Young *et al.*, 1972; Holliday *et al.*, 1974; Young *et al.*, 1975).

Nomura *et al.* (1969) obtained recordings of heartrate for one full 24h period from a 600g rainbow trout in a hatchery pond together with shoals of younger fish. The heartrate was fairly uniform with no clear diurnal rhythm; their recordings did not extend beyond the 24h recovery we have found necessary before normal behaviour can be observed. At least 3 or 4 days' data are required for analysis of rhythms and preferably much more for time series analysis techniques.

It is interesting to note that when fish D was away from its normal home range the heart rate was high and no circadian rhythm was apparent. Kanwisher *et al.* (1974) noted a similar phenomenon in Atlantic salmon recently brought into their observation tanks.

Kneis and Siegmund (1976) describe the results of long term monitoring of locomotor activity and heart rate in carp (*Cyprinus carpio* L). They used wire tethers attached to individual fish in aquaria for monitoring the ECG. They showed a correlation between circadian rhythms of heart rate and swimming activity. Both in summer and winter the heart rate and activity were higher during the day than at night, but in spring and autumn this was reversed. This is similar to phase shifts in locomotor activity reported in a number of teleosts. (Andreasson, 1969; Muller, 1969). No such reversal has been seen in Airthrey loch trout although fish D may be showing some evidence of it. Extensive heart rate telemetry work at different times of the year could prove very fruitful. Kneis and Siegmund (1976) also comment on the importance of the light-on and light-off stimuli, these being accompanied by the most pronounced changes in the heart rate and activity. The above curve fitting to the trout data confirms this view.

Priede and Young (1977) show by analysis of the increase in heart rate at dawn and decrease at dusk that there was evidence of anticipation of the solar time by about ½h in fish B. The values of   (Table 2; Fig. 5) for fish A, B and C lend support to such a hypothesis. Cardiac activity is concentrated in the day but there is a shift in the mode of the distribution to before noon. Examining Fig. 5 the transitions in the step function all slightly anticipate civil twilight.

Richardson and McCleave (1974) found no evidence of anticipatory changes in activity at dawn or dusk in juvenile Atlantic salmon (*Salmo salar*) but cite examples of such anticipation in sockeye salmon, (Nelson and Johnson, 1970) and Atlantic herring (Stickney, 1972). Anticipation could be explained by an endogenous rhythm, but no investigations have been undertaken of variations in heart rate of fish under constant illumination. An endogenous rhythm is not necessary to explain the apparent anticipation (Priede and Young, 1977).

No general conclusions can be drawn from the available

data for so few fish but heart rate telemetry seems a very
promising technique for the future. The heart rate is an
unequivocal event, accurately distinguishable in time.
Each heart beat interval can be precisely determined and
the large number (50,000-100,000 heart beats per day) of
such items of data readily lend themselves to statistical
analysis if automated logging systems are used.

The Relationship of Heart Rate to Physiology and Behaviour

In contrast to swimming activity or feeding the heart
rate is not an obvious proximately functional phenomenon.
Before assigning significance to measurements of cardiac
activity careful consideration must be given to the nature
and cause of change in heart rate.

Fig. 6 shows diagrammatically the relationships of the
heart to the rest of the animal. The cardiac output is a
function of stroke volume as well as rate so that large
fluctuations in blood flow and oxygen transport can take
place without any changes in heart rate.

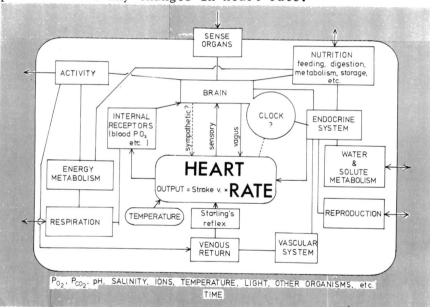

Fig. 6. Diagram showing some of the relationships of the heart rate
to the fish's internal physiology and the external environment.

Control of fish circulation is characterised by large
changes in cardiac stroke volume and small changes in
heart rate (Randall, 1968) but nevertheless there is a
correlation between heart rate and oxygen consumption
(Priede and Tytler, 1977) or swimming speed (Priede, 1974).
Mann (1965) observed daily fluctuations in oxygen consump-

tion in several freshwater fish and it is probable that
this would be reflected in the heart rate.

The heart is under direct neural control by the vagus
nerve which has a cholinergic inhibitory function (Randall,
1970). The heart intrinsically tends to accelerate in most
species in opposition to the vagal tonus (Labat, 1966).
The influence of the vagus is most evident in the 'réflexe
d'approche' (Labat, 1966) in which the heart beat is inhibited in response to a wide variety of external sensory
stimuli. This phenomenon of missing heart beats (Priede
and Young, 1977) may be used as an indicator of sensory
input. Fig. 7 shows cardiac inhibitions associated with
feeding activity in rainbow trout (*Salmo gairdneri*).
During field heart rate telemetry similar approach reflexes
were observed in response to passing boats and other disturbances (Priede and Young, 1977).

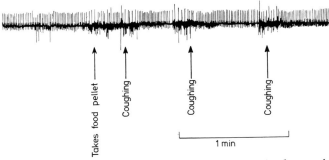

E.C.G. DURING FEEDING (Fish 27 15°C)

Fig. 7. The electrocardiogram of a rainbow trout (*Salmo gairdneri*).
Feeding in an aquarium. (Direct wire recording). The 'coughs' occur
during the swallowing process.

In trout as well as the cholinergic innervation there
are sympathetic andrenergic acceleratory fibres (Gannon
and Burnstock, 1969; Yamauchi and Burnstock, 1968; Gannon,
1971) but these have proved to be absent in plaice (*Pleuronectes platessa*) (Cobb and Santer, 1972). There are a number of sensory afferent fibres from the heart which mediate
some cardiac control reflexes (Laurent, 1962).

Chemoreceptors and mechanoreceptors in the vascular and
respiratory system therefore feed into the brain by afferent nerves and the heart is controlled in relation to the
requirements of respiratory gas transport etc. (Satchell,
1971).

Priede (1974) showed that heart rate changes during
exercise could be quite well regulated even after section
of the vagus nerves. Labat (1966) had investigated the
effects of vagotomy on a number of species of fish and
whilst some reflexes were abolished much cardiac function

is aneurally controlled. The heart has beta adrenergic
receptors (Falck *et al.*, 1966) and catecholamines have a
chronotropic and inotropic effect on the heart. Nakano
and Tomlinson (1967) found increase in the level of circ-
ulating catecholamines during exercise which could account
for the increase in heart rate.

Catecholamines also affect the rest of the circulatory
system. Randall and Stevens (1967) demonstrated that adrenaline could increase heart rate even after blocking of the
beta receptors. It is assumed that this effect is mediated
by alpha receptors in the blood vessels giving rise to
vasodilation in the gills and vasoconstriction in the sys-
temic circulation (Richards and Fromm 1969; Wood and Shelton, 1975). The heart responds to increase in venous return in accordance with Starling's law by an increase in
both rate and stroke volume (Bennion, 1968). In addition
to the effect of vaso-active agents the venous return can
be increased by mechanical pumping effects during swimming
movements (Priede, 1975).

The relationship between the heart and the endocrine
system is very complex since adrenaline for example has
important non-circulatory functions. Adrenaline influences
carbohydrate metabolism antagonising the effect of insulin
and promoting the release of glucose. Meier and Burns
(1976) review circadian rhythms in lipid regulation and
show that adrenaline and non-adrenaline stimulate fat
mobilization from adipocytes. They demonstrate a daily
rhythm of lipogenesis in killifish (*Fundulus grandis*);
this follows rhythms of food consumption but persists
even in starved fish. Hormone fluctuations associated
with this would also have a secondary effect on the heart
either directly or through vasoconstrictive or dilator
effects.

A number of hormones are involved in control of salt
and water balance. These are discussed elsewhere in the
symposium but neurohypophysial hormones (isotocin and
arginine vasotocin) have been shown to cause vasoconstriction in the gill lamellae whereas adrenaline causes vasodilation (Rankin and Maetz, 1971). These hormones are
involved in the control of filtration rates in the kidney
and electrolyte exchange in the gills (Maetz and Rankin,
1969). The osmoregulatory function is probably of primary
importance but effects on the heart may be noticed.

Randall (1970) lists various substances which can affect
the circulation and it is probable that endocrine changes
associated with the reproductive cycle could influence the
heart rate.

In Fig. 6 a hypothetical endogenous clock is shown,
probably associated with the neuroendocrine system. It is
questionable whether the heart would be directly influenced by this but the heart would be linked to it by a large
number of direct or indirect pathways involving rhythmic

phenomena in every aspect of the animal's physiology.
     Temperature is a very important ecological parameter
and can influence the heart in several ways. If a fish
senses a temperature difference this may be accompanied
by a 'réflexe d'approche' inhibition as for any other sensory stimulus.
     The body temperature of the fish determines the rate
of all metabolic processes. Therefore the heart rate must
change in order to accommodate associated changes in demands of respiratory gas transport etc. This would be mediated by the control mechanisms outlined above but the
functional characteristics of these pathways are themselves
influenced by temperature. Whether adrenaline increases
the rate of an isolated trout heart depends on the temperature (Bennion, 1968). Laffont and Labat (1966) showed
that intracardiac injection of adrenaline in the carp
(*Cyprinus carpio*) causes bradycardia at low temperatures
and tachycardia at high temperatures. Priede (1974) also
showed that vegotomy has different effects on trout at
different temperatures.
     Temperature directly affects the heart muscle itself.
Fig. 8 shows the effect of heating and cooling on the ECG
of a plaice in an aquarium.

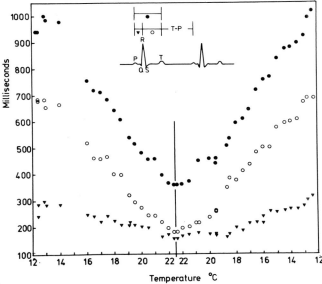

Fig. 8. The effect of heating and cooling on the ECG of a plaice
(*Pleuronectes platessa*) in an aquarium. The heating and cooling cycle
took approximately 3 hours. The diagram of the ECG indicates the
P-R, R-T and P-T intervals which are plotted against the temperature
scale. The maximum possible heart rate is when the T-P interval is
zero.

Labat (1966) shows similar curves for a number of other
species. The period between atrial depolarization (P wave)
and ventricular repolarization (T wave) is inversely proportional to temperature and acts as a limit on the maximum heart rate. (Wardle and Kanwisher, 1974). Priede and
Tytler (1977) show that knowledge of the maximum heart
rate is important in using heart rate as a measure of metabolic rate. If the full ECG is telemetered as in the
Kanwisher tag (Kanwisher *et al.*, 1974) the temperature of
the fish can be directly estimated from the P to T interval. Obtaining a clear enough signal in practice would
be difficult. This may be a useful additional technique
in the study of diurnal migrations of fish through temperature gradients. Differences in PO_2 in different layers
of water would also influence the heart rate as do a wide
range of environmental factors (Labat, 1966; Randall and
Smith, 1967).

The most obvious increases in heart rate must generally
be associated with activity, notably feeding. Many prey
species such as *Daphnia* (Haney and Hall, 1975) exhibit
diel migrations whilst grazing on algae. Thus apart from
considerations of light for visibility the activity cycle
of fish may also be influenced by the fact that some prey
are only available at certain times of the day. Swift
(1964) however showed that the activity cycles of trout
are not necessarily influenced by the time of feeding.

It has been shown that feeding activity accounts for
a very small part of the fish's total respiration (Warren
and Davis, 1967; Priede, 1977b). Metabolism during diges-
tion attributable to specific dynamic action is far greater
and persists for some time after a meal while the fish is
at rest (Beamish, 1974). Priede (1973) found that a trout
with a full stomach has an elevated heart rate which declines as the gut is evacuated. High heart rates can be
expected up to 12 or more hours after a meal and this
would be superimposed on other metabolic cycles such as
those of lipid deposition.

The heart rate is sensitive to innumerable influences
and it is hardly surprising that even minor stress such
as when the fish is displaced from its normal home range
results in high heart rates (Fish D).

We no longer regard the heart as the seat of the soul
or vital spirit. However it is still very much at the
centre of the animal's physiology and we can regard it as
a convenient transducer the output of which we can now
monitor throughout the fish's normal existence in the natural environment. Such measurements of heart rate can lead
to a fundamental reappraisal of our understanding of the
mechanisms of survival of fish (Priede, 1977a).

Acknowledgements

This work was supported by N.E.R.C. I thank Prof.

F.G.T. Holliday who initiated the project and numerous
assistants who made the practical work possible. I am indebted to Mr. A.H. Young and Mr. J. Wiewiorka who designed
the equipment and Mr. A. Lucas and Mr. R. Duthie for so
efficiently preparing the illustrations.

References

Andreasson, S. (1969). *Oikos* **20**, 78-94.
Beamish, F.W.H. (1974). *J. Fish. Res. Bd. Canada.* **31**, 1763-1769.
Bennion, G.R. (1968). The control of the function of the heart in
  teleost fish. M.Sc. Thesis. University of British Columbia cited
  by Randall, 1970.
Cobb, J.L.S. and Santer, R.M. (1972). *J. Physiol.* **222**, 42-43.
Falck, B., Mecklenburg, C. von, Myhrberg, H. and Persson, H. (1966).
  Acta physiol. scand. **68**, 64-71.
Frank, T.H. (1968). *IEEE Trans. Biomedical engineering.* BME **15**, 111-
  114.
Gannon, B.J. (1971). *Comp. Gen. Pharmacol.* **2**, 175-183.
Gannon, B.J. and Burnstock, G. (1969). *Comp. Biochem. Physiol.* **29**,
  763-773.
Greer-Walker, M., Harden Jones, F.R. and Arnold, G.P. (1977). *J. Cons.
  int. explor. Mer.* **38**, (in press).
Greer-Walker, M., Mitson, R.B. and Storeton-West, T. (1971). *Nature,*
  Lond. **229**, 196-198.
Haney, J.F. and Hall, D.J. (1975). *Arch. Hydrobiol.* **75**, 413-441.
Hawkins, A.D., MacLennan, D.N., Urquhart, G.G. and Robb, C. (1974).
  J. Fish Biol. **6**, 225-236.
Holliday, F.G.T., Tytler, P. and Young, A.H. (1974). *Proc. R. Soc.
  Edinb. B.* **74**, 315-331.
Kanwisher, J., Lawson, K., and Sundness, G. (1974). *Fish Bull. (U.S.)*
  72, 251-255.
Kelso, J.R.M. (1976). *J. Fish. Res. Bd. Can.* **33**, 42-53.
Kneis, P. and Siegmund, R. (1976). *Experientia:* **32**, 474-476.
Labat, R. (1966). *Ann. Limnol.* **2**, 1-175.
Laffont, J. and Labat, R. (1966). *J. Physiol. Paris* **58**, 351-355.
Laurent, P. (1962). *Archs. Anat. microsc. Morph. exp.* **51**, 337-458.
Maetz, J. and Rankin, J.C. (1969). *Colloques Internationaux du Centre
  National de la Recherche Scientifique,* **177**, 45-54.
Mann, K.H. (1965). *J. Anim. Ecol.* **34**, 253-275.
Mardia, K.V. (1972). *Statistics of Directional data.* Academic Press,
  London and New York.
Meier, A.H. and Burns, J.T. (1976). *Amer. Zool.* **16**, 649-659.
Müller, K. (1969). *Aquilo, Ser. Zool.* **8**, 50-62.
Müller, K. and Schreiber, K. (1967). *Oikos* **18**, 135-136.
Nakano, T. and Tomlinson, N. (1967). *J. Fish. Res. Bd. Can.* **24**, 1701-
  1715.
Nelson, D.R. and Johnson, R.H. (1970). *Copeia,* **1970**, 732-739.
Nomura, S. and Ibaraki, T. (1969). *Jap. J. Vet. Sci.* **31**, 135-147.
Nomura, S., Ibaraki, T., Hirose, H. and Shirahata, S. (1972). *Bull.
  Jap. Soc. scient. Fish.* **38**, 1105-1117.
Priede, I.G. (1973). The physiology of circulation during swimming
  activity in Rainbow trout. PhD Thesis. University of Stirling.

Priede, I.G. (1974). *J. exp. Biol.* **16**, 446-473.
Priede, I.G. (1975). *J. Zool., Lond.* **175**, 39-52.
Priede, I.G. (1977a). *Nature (Lond.)* **267**, 610-611.
Priede, I.G. (1977b). The respiratory metabolism of brown trout (*Salmo trutta* L.) in a Scottish loch measured by means of heart rate telemetry. (MS)
Priede, I.G. and Tytler, P. (1977). *J. Fish. Biol.* **10**, 231-242.
Priede, I.G. and Young, A.H. (1977). *J. Fish Biol.* **10**, 299-318.
Randall, D.J. (1968). *Amer. Zool.* **8**, 179-189.
Randall, D.J. (1970). *In: Fish Physiology* (W.S. Hoar and D.J. Randall, eds.), vol. 4. pp.133-172. Academic Press, New York.
Randall, D.J. and Smith, J.C. (1967). *Physiol. zool.* **40**, 104-113.
Randall, D.J. and Stevens, E.D. (1967). *Comp. Biochem. Physiol.* **21**, 415-424.
Rankin, J.C. and Maetz, J. (1971). *J. Endocrinol.* **51**, 621-635.
Richards, F.B. and Fromm, P.O. (1969). *Comp. Biochem. Physiol.* **29**, 1063-1070.
Richardson, N.E. and McCleave, J.D. (1974). *Biol. Bull.* **147**, 422-432.
Roberts, M.G., Wright, D.E. and Savage, G.E. (1973). *Comp. Biochem. Physiol.* **44A**, 665-668.
Satchell, G.H. (1971). *Circulation in Fishes.* Cambridge Monographs in Experimental Biology, No. 18. Cambridge University Press.
Spencer, W.P. (1939). *Ohio J. Sci.* **39**, 119-132.
Stasko, A.B. (1975). *Underwater Biotelemetry, an Annotated Bibliography.* Dept. of the Environment Fisheries and Marine Service Research and Development Directorate. Technical Report No. 534. St. Andrews. New Brunswick.
Stasko, A.B. and Polar, S.M. (1973). *J. Fish. Res. Bd. Can.* **30**, 119-121.
Stickney, A.P. (1972). *Ecology* **53**, 438-445.
Swift, D.R. (1964). *J. Fish. Res. Bd. Can.* **21**, 133-138.
Tesch, F.W. (1972). *Helgolander wiss. Meeresunters* **23**, 165-183.
Trefethen, P.S. (1956). *Spec. scient. Rep. U.S. Fish. Widl. Serv.* **179**, 11pp.
Wardle, C.S. and Kanwisher, J.W. (1974). *Mar. Behav. Physiol.* **2**, 311-324.
Warden, R.L. and Lorio, W.J. (1975). *Trans. Amer. Fish. Soc.* **104**, 696-702.
Warren, C.E. and Davis, G.E. (1967). *In: The Biological Basis of Freshwater Fish Production.* (S.D. Gerking, ed.), Blackwell, Oxford.
Wood, C.M. and Shelton, G. (1975). *J. exp. biol.* **63**, 505-524.
Yamauchi, A. and Burnstock, G. (1968). *J. comp. Neurol.* **132**, 567-588.
Young, A.H., Tytler, P., Holliday, F.G.T. and MacFarlane, A. (1972). *J. Fish. Biol.* **4**, 57-65.
Young, A.H., Tytler, P. and Holliday, F.G.T. (1975). *Proc. R. Soc. Edinb. B,* **75**, 145-155.
Young, A.H. and Wiewiorka, J. (1975). *Underwater Telemetry Newsletter* **5**, 10-13.

DEVELOPMENTAL CHANGES IN THE ACTIVITY RHYTHMS
OF THE PLAICE (*PLEURONECTES PLATESSA L.*)

R.N. GIBSON AND J.H.S. BLAXTER

Dunstaffnage Marine Research Laboratory, Oban, Scotland
and

S.J. de GROOT

Netherlands Institute for Fishery Investigations,
IJmuiden, The Netherlands.

Introduction

During its life history the plaice undergoes several
changes in its distribution and ecology. The eggs and
larvae are planktonic. As the larvae develop they begin
to swim actively and undergo diurnal changes in their
vertical distribution (Ryland, 1964). At metamorphosis
the young fish settle on the bottom and migrate shore-
wards onto sandy beaches. Once this onshore migration is
complete, the fish move up and down the beach with the
tide to feed. In this intertidal phase the movements of
the fish on the beach are well known (Gibson, 1973a;
Kuipers, 1973; Lockwood, 1974) and experimental studies
have shown that such fish possess an endogenous tidally-
phased activity rhythm (Gibson, 1973b, 1976). An attempt
has recently been made to assess the role that this rhythm
plays in the movements of fish in the sea (Gibson, 1975).

In their first winter the O-group fish move into deeper
water and in some areas may return the following spring to
continue their tidal movements. As growth continues the
fish move progressively offshore and at any given time
there is a direct relationship between body length and
water depth generally known as Heincke's law (Wimpenny,
1953).

The activity patterns of adult fish from deeper water
have been described by Verheijen and de Groot (1967) and
de Groot (1971). Such fish, in the laboratory, are most
active off the bottom at night; during the day they are
less active and remain on the bottom.

Throughout development, therefore, the fishes behaviour
undergoes several changes to enable them to adapt to their
new environment. The results presented in this paper represent an attempt to give a more detailed account of the
ontogenetic changes in locomotor activity patterns, to dis-
cover the environmental factors responsible for bringing

about these changes and to determine whether the activity
patterns described are endogenous or simply direct responses to the environmental conditions prevailing in
each habitat.

Materials and Methods

Methods of Recording Activity

Larvae 6-10 mm The activity of plaice larvae was monitored by thermistors at the surface of a vertical perspex
tube 4.5 cm in diameter and 120 cm high contained in a
light-tight box. Full details of the apparatus are given
by Blaxter (1973). Experiments lasting from 1-2 days were
performed in light:dark cycles (LD) of natural light and
in continuous dim light (LL 0.2 lx at surface) or continuous darkness (DD) after holding for at least 10 days in
a LD regime.

Juveniles 20-80 mm Activity was recorded in the laboratory in annular tanks using an infra-red light beam and
photocell as the detection device. This apparatus is described in greater detail in an earlier paper (Gibson,
1973b). A slow flow of water was provided throughout the
experiment and recordings were made in both DD and natural
LD cycles with light coming from a large N.E. facing window.

   Two series of experiments were conducted with juveniles
50-80 mm in length to test the hypothesis that cycles of
light and hydrostatic pressure were capable of entraining
a tidal rhythm. In the first series, fish that had been
kept in DD in the laboratory for 2-3 weeks were placed
in a water- and light-tight wooden box (90 x 45 x 45 cm)
anchored to the sea-bed near the laboratory at low water
mark of neap tides. Baffled tubes near the top of the
box enabled water to circulate through the box without
allowing light to enter. At low tide the box contained
water 35 cm deep. Fish were transferred from the laboratory to the box, held there for periods varying from 6-53
tidal cycles, and fed at irregular intervals. They were
then removed and their activity recorded in DD.

   In the second series of experiments, wild fish which
had been kept in the dark for 2-3 weeks prior to experimentation, were subjected to pressure cycles in a sealed
annular tank in which their movements could be continuously monitored. The tank was connected to a tide machine
capable of producing sinusoidal pressure cycles of varying
amplitude and period together with a continuous flow of
water. For the purpose of this experiment the tidal period
was set at 12.6 h with a range of 1.5 m. Individual fish
were put into the pressure tank in DD and subjected to
six pressure cycles after which the tide machine was
switched off and activity recorded for a further 50 h at
'low tide' pressure. Six pressure cycles were chosen

because it was found that exposure to five tidal cycles
in the shore box was sufficient to entrain a tidal rhythm
(see below).

The same pressure chamber was used to record the activity of individual fish subjected to cycles of light and
pressure combined. In this case illumination was provided
by a circular 32 W fluorescent tube above the chamber
which gave a light intensity of 450 lx. The pressure
regime was the same as the previous experiment, but in
addition, the light was switched off from 2100 to 0400
each night and during the light period a further three
hours of darkness was given 1 h later each day (1100-1400,
1200-1500 and 1300-1600) to simulate a reduction of light
intensity caused by an increase in water depth at high
tide. Each 'high tide' thus occurred in darkness. After
6 light and pressure cycles the tide machine and light
were switched off and activity recorded for a further
50 h.

The same light:dark regime was used to test the effect
of LD cycles in the absence of pressure variations. For
this experiment, activity was recorded in the open annular
tanks described above, using the same 32 W fluorescent
tubes to give a light intensity of 450 lx.

Adults 20-30 cm Activity was recorded in an oval annular
tank (4.2 x 2.4 x 1.0 m deep). The method used (de Groot
and Schuyf 1967; Schuyf and de Groot, 1971), depends on
the detection of the induction voltage generated in a
fixed coil when a small magnet attached to the fish crosses
the coil. In all, four coils were used, two just above the
water level and two buried in the sand covering the tank
bottom. The coils detect the passage of fish at a distance
of 10 and 50 cm, and two types of activity were defined.
Surface activity (detected by the upper coils) includes
all activity in which the fish is <50 cm from the surface
(>50 cm from the bottom) and bottom activity (detected
by the lower coils) in which the fish is 10 cm or less
from the bottom.

Experiments were performed in LD, LL and DD. The light
source consisted of two fluorescent tubes (Philips TLM
40/W33RS).

Origin of Experimental Fish

Larvae The larvae used in the experiments were reared in
the laboratory from artificially fertilised eggs and fed
on *Artemia* nauplii except during an experiment.

Juveniles Initially an attempt was made to collect fish
that had just metamorphosed and settled on the bottom
from depths of 6-10 m off a sandy beach. These fish had
not as yet moved into the intertidal zone and were therefore considered to be at a stage intermediate between the
larvae and the juveniles already in shallow water. Although

some individuals were collected at this stage, difficul-
ties were experienced in finding sufficient numbers for
an extended period, as well as catching them in a healthy
condition. Since the purpose of using fish at this stage
in development was to determine their activity pattern
before entering the intertidal phase of their life history,
it was considered that reared fish might serve this purpose
equally well. Reared fish have the advantage of being
readily available, but do suffer from the disadvantage
that the relationship of their behaviour to that of wild
individuals is not known. Once the fish had become intertidal, however, they were easily obtained in good condition from the beach.

Adults Adult fish were caught off the Dutch coast by inshore research vessels or shrimp trawlers in depths of
approximately 20 m. The fish were kept for 2-5 weeks before
being used in the experiments and any that showed signs
of damage or did not feed were discarded.

Results

Activity

Larvae As judged by recording the activity at the surface
of the vertical migration tube, young larvae showed a marked rise to the surface at dusk. They remained there
throughout the night and moved away from the surface at
dawn (Fig. 1a). Older larvae, 40 or more days post-hatching, which were starting to settle on the bottom showed
a much lower general level of activity (see Blaxter and
Staines, 1971). The present experiments demonstrate that
vertical migration also disappears at this stage. Metamorphosis usually occurred at 50-60 days post-hatching;
the fish then spent almost all the time on the bottom of
the rearing tanks. When larvae of different ages were
held in a natural or artificial LD cycle for at least
10 days and then submitted to DD (Fig. 1b) or LL (Fig. 1c)
no endogenous rhythm was observed except perhaps a slight
one at 13 days post-hatching.

Juveniles: wild fish before becoming intertidal It only
proved possible to do one short term experiment with such
fish which measured 19-20 mm in length. The results suggested, however, that in DD the fish are most active at night
and that such activity is not related to the tides (Fig.
2a). The nocturnal activity pattern seemed to have two
peaks, a large one in the early morning (0100-0500) and a
smaller one in the late evening (2100-2200).

Laboratory-reared juveniles 30-40 mm Two sets of experiments were performed with these fish, one in DD and another
in LD. In DD the fish exhibited a bimodal activity pattern
with a peak in the late afternoon (1600-1800) and another
in the early morning (0100-0400) (Fig. 2b). In LD the

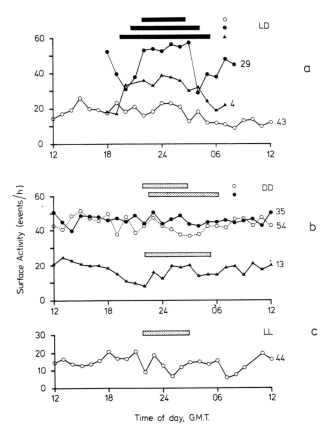

Fig. 1. Activity of plaice larvae at the surface of the vertical migration tube at different times. Fig. 1a shows effect of LD cycles of natural light, Fig. 1b the effect of DD and Fig. 1c the effect of LL (dim light), both after at least 10 days of LD. The black bars in Fig. 1a show the duration of the dark period between civil twilights and in Fig. 1b and c the times of darkness in the previous LD cycle are indicated by the shaded bars. The age of the plaice is given on the right of each graph in days post hatching. The temperature increased from about 10^0 to 14^0C during the growth of the larvae. The graphs in Fig. 1a represent the results from single nights; those in Fig. 1b,c, represent the mean result at each hour for two consecutive nights.

fish were also rhythmically active showing two peaks every 24 h, but in this case they appeared at 1900-2000 and 0500-0800 (Fig. 2c).

Wild juveniles 40-70 mm Wild fish from the beach also showed a rhythm with two peaks each day, but the peaks

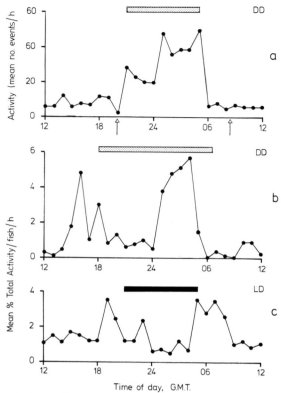

Fig. 2. The activity of 8.20 mm plaice collected in 6-10 m and re-corded in DD for 48 h. Activity is expressed as mean counts/h. The shaded bar above the graph indicates the time of the 'expected' dark period and the arrow the times of predicted high tide. (b) The activity pattern of 5 reared plaice in DD. Activity is expressed as a mean hourly percentage of the total 5 day experiment. (c) As b, but activity of 8 fish recorded in LD. The dark period is shown by the black bar.

are phased with the tides. Such a rhythm persists in DD in the laboratory for only 2-3 cycles (Gibson, 1973b, 1975, 1976). Analysis of the results of an experiment run over a 15-day semi-lunar cycle, with fish being replaced every two days, demonstrated that the initial rhythm has a circatidal rather than circadian character (Fig. 3a,b). This circatidal rhythm rapidly reverts to a bimodal circadian rhythm if the fish are kept in the laboratory for several days (Fig. 3c,d). The two peaks of the bimodal circadian rhythm are unequal in height and the general pattern of activity resembles that shown by immediately post-metamorphic wild fish (Fig. 2a) and by laboratory-

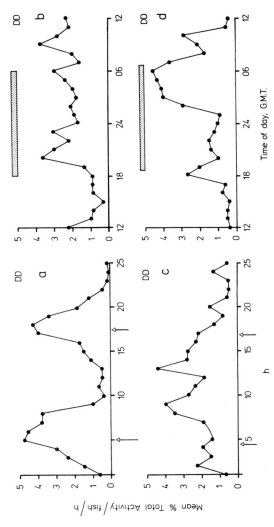

Fig. 3a. 25 h (twice tidal) average of a 15-day experiment with wild fish. The fish were replaced every 2 days and the graph shows the circatidal nature of the rhythm. (b) 24 h average of the data in (a). (c,d) as (a,b) but using fish that had been kept in LD in the laboratory for 7 days before the experiment began. These two graphs (c,d) show the loss of the tidal rhythm and the presence of a bimodal circadian rhythm in such fish. Symbols as Fig. 2. Activity is expressed as the mean hourly percentage of the total. Data from Gibson (1973b).

reared individuals (Fig. 2b).

Adults Experiments were performed with adult plaice in
LD, LL and DD and the recording technique enabled activity on or close to and off the bottom to be distinguished.
In LD the fish were most active on or close to the bottom
during the day and there was a tendency for such activity
to be most intense near the beginning and end of the light
period (Fig. 4a). During the night, bottom activity was
greatly depressed and surface activity enhanced. The swimming behaviour of the fish thus seems to depend on the
light intensity; light favouring bottom activity and darkness surface activity. The experiments in continuous illumination confirmed this suggestion that light depresses
surface activity, but it also seems to reduce the overall
level of activity. Under these conditions the fish continue to show a rhythm in bottom activity but with only
a single peak during the few hours after midnight. There
was also a small rise in surface activity at the same
time (Fig. 4b). In DD the reverse occurs with bottom activity being relatively depressed and surface activity enhanced. As in LL, the rhythm is still clear, but in this
case it is most obvious in the surface activity (Fig. 4c).
Light intensity thus has a marked effect on the type of
locomotor behaviour shown by the fish, but in both LL and
DD a circadian rhythm persists.

In addition, the effect of darkness on swimming behaviour varies within the 24 h period. When given during the
normal light period (day) it suppresses bottom activity less
than during the normal dark period (night). It has little
effect on surface activity when given during the day, only
increasing this type of behaviour during the normal light.
These results suggest that there is a diurnal rhythm in the
responsiveness to darkness as well as a rhythm of swimming
activity. There appears to be no such rhythm in the responsiveness to light because it enhances bottom activity and
suppresses surface activity to an equal extent whatever
time of the day or night it is given (Fig. 5).

When comparing the activity patterns of the adults
with that of the juveniles, where the two types of activity were not separated, the total activity count of the
adults should be used. When compared in this way, the
main difference lies in the fact that the pattern shown
by juveniles in DD is bimodal rather than unimodal.

Entrainment of Tidal Rhythmicity

It seems likely that the circadian rhythm of the post-
metamorphic juveniles, laboratory-reared juveniles and
adults is entrained by light, but the question also arises
of how this circadian rhythm becomes tidal in character
once the young fish move into the intertidal zone. If it
could be demonstrated that some environmental factor was
capable of entraining a tidal rhythm in fish whose rhythm

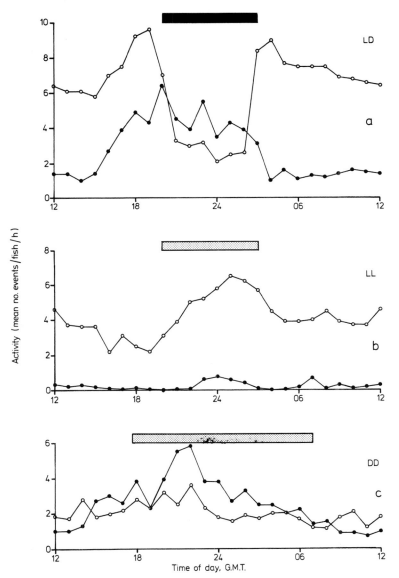

Fig. 4. 24 h averages of the activity of adult plaice (a) 10 fish over a 7-day period in LD. (b) same 10 fish as in (a) for a further 6 days in LL (approx. 40μWscm^{-2}). (c) 8 fish over a 6-day period in DD (approx. 8μWscm^{-2}) O surface activity, ● bottom activity. Other symbols as Fig. 2. Activity is expressed as the mean no. counts/fish/h.

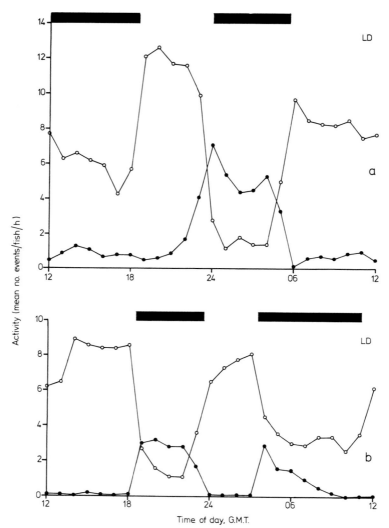

Fig. 5. 24 h averages of the activity of adult plaice. (a) 8 fish over a 7-day period in LD with an extra dark period during the day (1200-1830). (b) 9 fish over an 11-day period in LD with a light period during the night (2330-0330) and an extended dark period. Symbols as Fig. 4.

was initially circadian, then the presence of tidal rhythmicity in intertidal fish could be explained. The most obvious, but not the only candidates for tidal zeitgebers are the cyclic variations in light intensity and hydro-

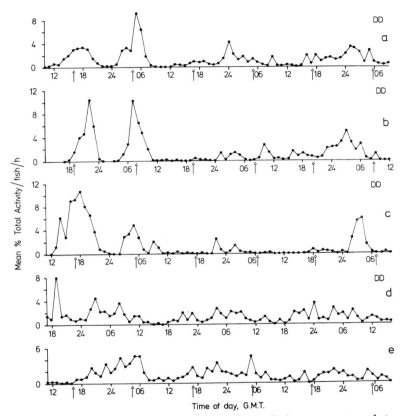

Fig. 6. The results of experiments in which fish were exposed to hydrostatic pressure cycles in the field. (a) After exposure to 5 tidal cycles of 1.1-1.4 m range in DD. (b) After exposure to 53 tidal cycles in DD. (c) After exposure to 32 tidal cycles in LD. (d) An example of the activity pattern of fish before being placed in the shore box, control experiment for c. (e) After exposure to 30 tidal cycles on a floating raft. All activity was recorded in DD and is expressed as the mean % of total activity/fish/h. 5 fish in each experiment. The arrows show the predicted time of high tide.

static pressure associated with the ebb and flow of the tides. The experiments in which the fish were exposed to natural and laboratory light and pressure cycles were designed to investigate this problem.

The results of the experiments in which the fish were

exposed to natural hydrostatic pressure cycles are given
in Fig. 6. Fig. 6a shows the pattern of activity after
the fish had been in the box for 5 tidal cycles. After
this treatment the fish showed two peaks of activity in
the first 24 h which were close to the expected times of
the high tides. Exposure to 53 tidal cycles did not markedly improve the precision or persistence of the rhythm
(Fig. 6b). LD cycles in combination with pressure cycles,
obtained by using an open box, also do not appear to enhance entrainment (Fig. 6c). The activity pattern after
treatment may be compared with the pattern beforehand
in which no tidal rhythmicity is apparent (Fig. 6d). The
control experiment with fish held on a floating raft also
produced no evidence of entrainment of a tidal rhythm
(Fig. 6e). In some of the activity patterns exhibited by
fish before being placed in the shore box there was a
suggestion of a weak circadian rhythm in that the fish
were more active during the subjective night (Fig. 6d).
A similar phenomenon was also noticed in the control experiment (Fig. 6e). In those experiments where the fish
were subjected to pressure cycles, the initial tidal
peaks rapidly gave way to a large peak in the early morning (Fig. 6a,b,c). This shift of tidal peaks was also
found in earlier experiments (Gibson, 1973b, 1976), and
it is considered that this shift represents a spontaneous
reversion from a circatidal to a circadian rhythm.
    The second series of entrainment experiments in the
laboratory allowed activity to be recorded while the
fish were exposed to light and pressure cycles. Fig. 7a
demonstrates that the fishes activity reflects the pressure cycles to a large extent, but that, in general,
their activity is greatest as the pressure starts to
fall after 'high tide'. Fig. 9a which represents the
average pattern over a 25 h (twice tidal) period demonstrates this point more clearly, and it is noticeable
that there is a marked similarity between this average
pattern and that obtained from freshly caught wild fish
(compare Figs. 3a and 9a). When the pressure was held
constant, there was no clear evidence of an entrained
rhythm although activity was higher close to the expected
times of the first and third high waters (Fig. 8a). This
pattern results in the 25 h average having only one noticeable area of high activity at the expected time (Fig. 9b).
In LD cycles the fish were generally more active during
the dark period than in the light (Figs. 7b,9c) but in
the final 50 h of darkness there was no evidence of an
entrained circa-tidal rhythm, although during the subjective night activity was higher than average suggesting
that a weak nocturnal circadian rhythm was present (Figs.
8d,9d). The results of the experiment in which the fish
were subjected to a combination of light and pressure
cycles is illustrated in Fig. 7c and diffuse peaks of

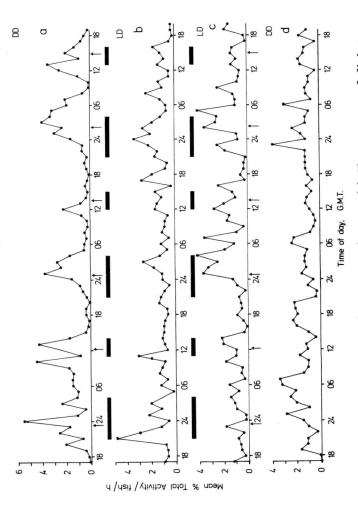

Fig. 7. The results of the laboratory entrainment experiments. (a) The response of fish to pressure cycles of 1.5 m range at 12.6 h periods in DD. 8 fish. (b) The response to light:dark cycles with the short dark period being delayed by 1 h. each day. 9 fish. (c) The response to a combination of pressure and LD cycles. 7 fish. (d) Control experiment in DD in the absence of pressure cycles. 5 fish. The arrows show the times of 'high tide' and the black bars the dark period.

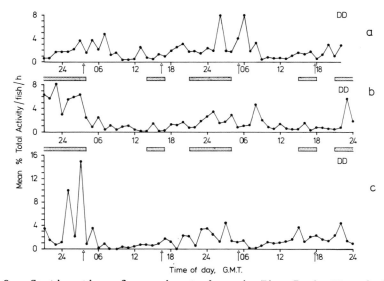

Fig. 8a. Continuation of experiment shown in Fig. 7a in DD and the absence of pressure cycles. (b) Continuation of the experiment shown in Fig. 7b in DD. (c) Continuation of the experiment shown in Fig. 7c in DD and the absence of pressure cycles. The arrows show the 'expected high tides' and the shaded bars the 'expected dark periods'.

activity can be recognised at the times when the pressure was high and the lights were off. The 25 h average of the experiment confirmed this impression (Fig. 9c). The activity pattern recorded in DD in the absence of pressure cycles again shows no obvious entrainment of rhythm of approximately tidal period, but the pattern is similar to that of the previous experiment (Fig. 9f) in that activity is highest during the night. These results confirm those where the fish were held on the shore in an open box, in that LD cycles do not enhance entrainment.

Summarising these laboratory experiments, it seems that although the fish do respond to light and pressure cycles, any resulting entrained rhythm that may be present is not obvious and does not compare in precision with that obtained from fish subjected to natural light and pressure cycles by returning them to the shore.

Discussion

Plaice larvae only exhibit a rhythmic pattern of vertical migration in the presence of LD cycles; in LL or DD no clear pattern is evident, suggesting that larval swimming behaviour is directly controlled by the environment rather than by internal factors. Some time after metamorphosis, however, the activity patterns recorded in the laboratory show clear signs of rhythmicity both in LD and

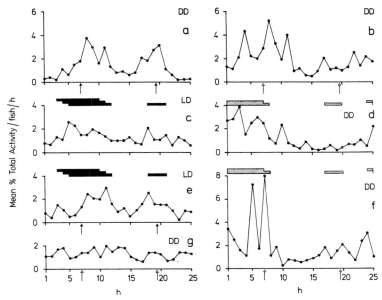

Fig. 9. 25 h averages of the data in Figs. 7 and 8. (a) From Fig. 7a (b) From Fig. 8a (c) From Fig. 7b (d) From Fig. 8b (e) From Fig. 7c (f) From Fig. 8c (g) From Fig. 7d. Actual dark periods are shown as black bars and expected dark periods as shaded bars, the times of actual 'high waters' by solid arrows and expected 'high waters' by open arrows. 25 h averages were calculated in order to detect the presence of circatidal rhythmicity.

constant conditions, although the exact pattern depends on the origin of the fish. In immediately post-metamorphic and laboratory-reared juveniles the rhythm has a circadian period and exhibits two peaks each 24 h both in LD and DD. The adult's circadian rhythm in LD also tends to be bimodal, but in LL and DD it is unimodal.

The circadian rhythm of very young juveniles is entrained to a circatidal rhythm once they become intertidal and the experiments designed to determine the environmental factors capable of such entrainment suggested that hydrostatic pressure cycles are at least partially responsible. There was a discrepancy in these experiments between the results obtained after subjecting the fish to pressure cycles in the laboratory and in the field. Although the fish responded to pressure cycles in the laboratory there was no obvious resulting entrained rhythm under these conditions, whereas well marked entrainment was noticed after the fish had been subjected to natural pressure cycles of comparable range for the same period of time.

One possible explanation of this discrepancy may lie
in the timing of the pressure cycles with respect to the
activity peaks of the endogenous circadian rhythm. In the
field experiments where entrainment was achieved after
exposure to only five tidal cycles, the high (neap) tides
occurred in the late evening and early morning and would
thus reinforce the two peaks of any residual bimodal circadian rhythm present in the experimental fish. In the
laboratory experiments, the times of the 'high tides'
were deliberately chosen so that they would not reinforce
any circadian rhythm that may have been present. Rapid
entrainment of a circatidal rhythm may therefore be achieved only when the tidal zeitgeber (pressure) is in phase
with the bimodal circadian rhythm. It is interesting to
note in this context that the fishes response to pressure
was most marked when the experimental 'high tides' coincided with the expected night (Fig. 7a). Apart from the
synchronisation of the pressure cycle with the circadian
rhythm, there is probably another factor of importance
in entraining circatidal rhythmicity and that is the magnitude of the pressure change. On spring tides (range 3 m),
the high waters occur at approximately midnight and noon.
They are thus out of phase with the two peaks of the circadian rhythm and yet under these conditions in the field,
entrainment of a tidal rhythm was also obtained after six
tidal cycles. Other laboratory experiments confirmed that
the amplitude of the pressure cycle is important, because
the fish responded only weakly to pressure cycles of 1 m
range and showed no response at all to cycles of 0.5 m
range.
    It is not known how the fish, which have no swimbladder,
perceive such pressure changes. It is clear that they can
detect pressure changes as larvae (Qasim, Rice and Knight-
Jones, 1963; Rice, 1964), juveniles (this paper) and
adults (Blaxter and Tytler, 1973). A suggestion has been
put forward by Morris and Kittleman (1967) however, that
since the otoliths of at least some fish possess piezo-
electric properties, such structures could act as pressure
receptors.
    Although some zeitgeber with a tidal period must be
present in the shallow water where the juveniles live,
the way it acts is difficult to understand. It is known
from field observations that the majority of fish in
their intertidal phase move up and down the shore in a
fairly well-defined band at a depth of 1-2 m (Gibson,
1973a). The majority of the population is therefore always
in approximately the same depth of water and hence not
subjected to the cyclical variations in the environmental
factor(s) likely to act as zeitgeber(s). Those fish which
do not migrate, on the other hand, will be subject to
pressure variations for example, the extent of which will
depend on the tidal range. It is conceivable that the

fading of the tidal rhythm observed in the laboratory may
also occur in the sea, and if, as seems likely, the endogenous tidal rhythm plays some part in controlling their
tidal migrations (Gibson, 1975), this fading may result
in a number of individuals failing to migrate. Failure to
migrate exposes the fish to pressure cycles which could
resynchronise their rhythm and allow them to relate their
activity to the tidal cycle once again. In some areas,
however, notably the Wadden Sea, virtually the whole population migrates into the intertidal zone (Kuipers, 1973),
which implies, if the above hypothesis is correct, that
their tidal rhythm is more persistent. There is some evidence to support this contention, because fish which migrate onto the extensive sand flats of the Wadden Sea do
have a more persistent tidal rhythm than those both of
adjacent beaches, where the migration is not as extensive,
and of the fish used in the experiments described here
(Gibson, personal observation).

The change from a tidal to a bimodal circadian rhythm
in the absence of a tidal zeitgeber has been described
in this and earlier papers (Gibson, 1973b, 1976) and it
is considered that such a change may also occur in nature
after the fish move offshore into deeper water. Work described here, and earlier by Verheijen and de Groot (1967)
and de Groot (1971), on the adults has demonstrated the
presence of circadian rhythmicity in such fish. In other
species too, the flounder (*Platichthys flesus*) for example,
young individuals living in tidal conditions possess a
tidally-phased rhythm (Gibson, 1976) whereas similar
stages living in the Baltic, where the tides are of less
importance, have a bimodal circadian rhythm (Muus, 1967).

How far these patterns of activity recorded in the
laboratory are representative of the behaviour of the
fish in the sea has yet, for the most part, to be determined. In general however, the type of rhythmic pattern
exhibited by fish at each stage in their life history is
appropriate to its particular environment. The planktonic
larvae show a vertical migration rhythm which becomes less
conspicuous as they develop and settle on the bottom. The
juveniles' activity rhythm is phased with the tides only
when they inhabit beaches with a marked tidal range, and
the level of the adults' activity, although greatly modified by prevailing light intensity, seems to be controlled
by an endogenous circadian rhythm. The laboratory activity
pattern of juveniles from tidal conditions is not completely representative of its behaviour in the sea (Gibson,
1975) however, and although there is some evidence to
confirm that the diurnal alternation in on- and off-
bottom activity of the adults observed in the laboratory
also occurs in the sea (de Groot, 1971), the function of
the nocturnal mid-water swimming still has to be fully
explained.

186   R.N. GIBSON ET AL.

References

Blaxter, J.H.S. (1973). *J. mar. biol. Ass. U.K.*, **53**, 635-647.
Blaxter, J.H.S. and Staines, M.E. (1971). *Proc. 4th Europ. mar. biol. Symp.* (D.J. Crisp, ed.), 467-485. Cambridge Univ. Press.
Blaxter, J.H.S. and Tytler, P. (1973). *Symp. Soc. exp. Biol.*, **26**, 417-443.
Gibson, R.N. (1973a). *J. exp. mar. Biol. Ecol.*, **12**, 79-102.
Gibson, R.N. (1973b). *Mar. Biol.*, **22**, 379-386.
Gibson, R.N. (1975). *Proc. 9th Europ. mar. biol. Symp.*, (H. Barnes, ed.), 13-28. Aberdeen Univ. Press.
Gibson, R.N. (1976). *In: Biological Rhythms in the Marine Environment.* Belle Baruch Library in Marine Science, No. 4, (P.J. DeCoursey, ed.), 199-213. Univ. S. Carolina Press, Columbia, S. Carolina.
Groot, S.J. de. (1971). *Neth. J. Sea Res.*, **5**, 121-196.
Groot, S.J. de. and Schuyf, A. (1967). *Experientia*, **23**, 574-576.
Kuipers, B. (1973). *Neth. J. Sea Res.*, **6**, 376-389.
Lockwood, S.J. (1974). *J. Fish Biol.*, **6**, 465-477.
Morris, R.W. and Kittleman, L.R. (1967). *Science, N.Y.*, **158**, 368-370.
Muus, B.J. (1967). *Meddr Danm. Fisk.-og Havunders.*, **5**, 1-316.
Qasim, S.Z., Rice, A.L. and Knight-Jones, E.W. (1963). *J. mar. biol. Ass. India*, **5**, 289-93.
Rice, A.L. (1964). *J. mar. biol. Ass. U.K.*, **44**, 163-175.
Ryland, J.S. (1964). *J. mar. biol. Ass. U.K.*, **44**, 343-364.
Schuyf, A. and de Groot, S.J. (1971). *J. Cons. perm. int. Explor. Mer.*, **34**, 127-132.
Verheijen, F.J. and de Groot, S.J. (1967). *Neth. J. Sea Res.*, **3**, 315-322.
Wimpenny, R.S. (1953). *The Plaice.* Buckland Lectures for 1949, Arnold and Co., London.

TIMES OF ANNUAL SPAWNING AND REPRODUCTIVE STRATEGIES IN AMAZONIAN FISHES

HORST O. SCHWASSMANN

*Department of Zoology, University of Florida,
Gainesville, Florida, 32611 U.S.A.
and
Instituto Nacional de Pesquisas da Amazonia, Manaus,
Amazonas, Brazil*

General Geographic and Climatic Features

With an area of more than 6×10^6 km^2, the South American Amazon basin is by far the largest river drainage system on earth (Soares, 1959). The Amazon is 6,571 km long when considering the Apurimac-Ucaiali as its headwaters (Soares, 1959) which makes it the second longest river on this planet, the Nile being a little longer. No other river, however, comes even close to the enormous water volume which the Amazon discharges into the Atlantic Ocean. Latest estimates give an average discharge of more than 200,000 m^3/sec (Oltman *et al.*, 1964), and this figure does not include the southern part of the delta, the Rio Pará, which consists of the large Tocantins and several other smaller tributaries. One of these, the Rio Guamá entering the Baía de Guajará at the city of Belém can be cited for comparison. The discharge of this small tributary was 2,000 m^3/sec during the rainy season (Egler and Schwassmann, 1962). The width of the Amazon varies considerably during the year, depending on high or low water level, and from place to place due to the existence of many alluvial islands and side arms. At the narrowest place, near Obidos, it is about 2 km wide; most commonly the distance between the two shores is from 10 to 20 km. The depth varies between 20 and 100 m, and the current's velocity is from 0.5 to 3 m/sec.

Only from an aeroplane is it possible to obtain some realistic picture of this enormous river system consisting of one or more river arms, floodplain (or varzea) lakes, ox-bow lakes, channels (furos), and flooded forests (igapó). The central basin consists of the equatorial lowland rainforest and is surrounded by the Andean foothills in the West, the Guyana shield in the North, the Central Brazilian shield in the South, and the Atlantic ocean and the North-Eastern Brazilian semi-arid scrubland

in the East. At the higher levels of the Guyana shield and
also in the Central Brazilian uplands, the rainforest gives
way to savanna where the separation of the year into wet
and dry seasons is very pronounced, while rainfall is more
evenly distributed in the central basin.

Average rainfall in this area is 2.5 m/annum which can
be converted to 15.0×10^{12} m^3 in one year, or 500,000 m^3/
sec. Since the average river discharge is about 2/5 of this
amount, one can assume that 3/5 or 300,000 m^3/sec is being
recycled as evaporation/precipitation.

The enormous amount of water is channelled through the
immense system of interconnecting bodies of predominantly
running waters and determines the characteristic prevailing landscape. According to their colour and degree of
turbidity, three major types of river water are recognized
(Sioli, 1965):

      1. Whitewater rivers of yellow-whitish turbidity
      caused by suspended sediment (Solimões, Madeira),
      2. Clearwater rivers, very transparent, sometimes
      with a yellowish tinge (Xingú, Tapajos),
      3. Blackwater rivers of dark-brown or black colour
      and low turbidity (Rio Negro).

General Observations about Amazonian Fishes

In this largest river basin there exists a rich fish
fauna with about 1,300 different species of mostly primary
freshwater fishes. How this great species diversity may
have come about, due to such factors as availability of
diverse habitats, lack of adverse glaciation effects,
predator pressure, etc., has been discussed by Lowe-
McConnell (1964, 1969, 1975), Géry (1969), and Roberts
(1972).

The Amazon basin and the adjoining river drainage areas
(Fig. 1) include some permanent and temporary headwater
interconnections, the 'ligacões de bacias', which most
probably served as pathways for range expansion by many
species of fish. The map also illustrates that the Amazon
basin has a much greater extension South than North of the
equator.

With respect to species distributions, it has been noted
that the fish faunas in the peripheral regions, mainly the
Guyana and Brazilian plateaux, show greater similarities
with each other than with those in the central Amazon
flood plain (Géry, 1969), but there are many species that
are found over the entire Amazon and adjacent basins as
geographically isolated and distinct populations, especially those inhabiting headwater streams (Fig. 1 and
Schwassmann, 1976). These species seem to exemplify the
process of allopatric speciation.

It is generally assumed that the richness in different
species is not paralleled by high population densities;
however, reliable information concerning fish standing

crops and yields in Amazonian waters is lacking. The noted
relatively low primary and secondary production is caused
by two main factors; a known paucity of available nutrients
in soil and water, and a high degree of turbidity in many
running waters which prevents penetration of sunlight necessary for photosynthesis.

Fig. 1. Map of the South American continent showing the major river
basins: Amazon drainage system (1); Orinoco (2); Guyana plateau rivers
Essequibo, Rupununi, etc. (3); Rio Paraguay - Paraná - La Plata system
(4); N-E Brazilian rivers, mainly Rio São Francisco (5); Magdalena
river basin (6); Some of the headwater interconnections (ligacões de
bacias) are marked by lower case letters: Cassiquiare between Orinoco
and Rio Negro (x); Rupununi-Rio Branco (y); Rio Guaporé - Rio Paraguay
(z); Tocantins - Rio Paraná (z^1). Capital letters indicate study sites
of populations of the gymnotoid *Gymnorhamphichthys* referred to later
in the text; small circles point to additional collecting sites of
this species (Nijssen, Isbrücker, and Géry, 1976).

Higher concentrations of algae, zooplankton, and fishes
are found in areas of standing water where sediments can
settle to the bottom, permitting light penetration. Examples are the varzea lakes and the interior of the island
of Marajó.

It can be expected that agricultural development in
this area will result in an enrichment in nutrients that
will be added to the water with the rainwater run-off.
Eutrophication might occur near regions of increased agricultural usage, thus increasing the potential productivity
of Amazonian waters.

Regardless of the manner in which agricultural and in-
dustrial development proceeds, one result will be a dis-
turbance of existing delicate ecological balances in the
aquatic environment. Because of their often highly specialized status, existing species might not be able to adapt
their physiology and behaviour sufficiently fast to the
altered conditions. To counteract any detrimental effects
on fish production, basic studies on the biology and behaviour of commercially important species have been initiated at the Instituto Nacional de Pesquisas da Amazonia,
Manaus, and data on growth rate, food requirements, times
of reproduction, timing and extent of migrations, and degree of tolerance to diverse environmental conditions,
are now accumulating. These basic studies must also include other not directly valuable species which either
serve as food organisms for the economically important
species, or which can be used as index organisms more
easily than the latter. Knowledge about the normal times
of reproduction and the environmental factors which determine the onset of spawning and the timing of the rhythm
of gonadal ripening is of primary importance for the ultimate success of planned aquaculture methods which will
certainly become of widespread practice in the future.
The success of such an approach of initial basic research
and eventual application of the achieved knowledge has
been well demonstrated in the case of the Brazilian North-
East which now produces many thousands of tons of fish in
a system of ponds and reservoirs.

Seasonal Timing of Reproduction and Problems of the Tropics

Reproduction in most species of teleost fish is annually
periodic. Annual spawning usually occurs at a time of year
when environmental conditions are favourable for survival
of the offspring, especially with regard to food supply
and availability of space.

Concerning the advance timing of spawning and related
activities, it is usually assumed that an endogenous physiological rhythm of gonadal maturation, with demonstrated
pituitary-gonadal interactions, is regulated in time of
year by the action of certain environmental variables.
Of these, the systematically changing duration of daylight

(photoperiod or photofraction) is considered the most
reliable, and photoperiodic control of annual reproduction
and sometimes migrations has been demonstrated in several
species of bony fishes of the temperate zones (Schwassmann,
1971).

In the tropics, especially near the equator where the
annual change of daylength is very small or non-existent,
control of seasonal reproduction by means of photoperiod
can be ruled out. Here, in the majority of cases, there
seems to exist a coincidence of migrations and spawning
with the earliest extensive flooding of wide areas due
mostly to the rising water level during the rainy season.
Since there are many instances of these activities occurring within only days after the earliest heavy rains, this
factor - flooding - cannot be held responsible as an advance timing mechanism for the rhythm of gonadal ripening.
Whether such anticipatory timing does exist, or whether
perhaps a relatively long-lasting phase of spawning-readi-
ness is simply terminated in a trigger fashion by the
earliest flooding stage, or by extensive rains, has not
been satisfactorily determined.

The Seasonal Cycle in the Different Main Biotopes

Floodplain lakes, Varzea and Igapó

Rainfall, run-off, and floods In the peripheral regions
of the Amazon basin we find a clear separation of the year
into a rainy and a dry season while rainfall in the central
lowlands is more evenly distributed. On the northern Guyana
shield, and markedly also in the upper Rio Negro drainage,
the heaviest rains fall between April and August. On the
southern Brazilian shield these same months are normally
dry and October to April is the time of heavy rains. As
was already mentioned, no really dry and wet seasons occur
in the main valley but the early months of the year, January to May, have about twice as much precipitation as the
rest of the year (Fig. 2).

These seasonal differences in precipitation affect the
volume of run-off and the water discharge of the main
rivers, accompanied by considerable changes in the water
level. At Manaus, near the confluence of Solimões and Rio
Negro, the annual difference in river level is approximately 10 m (Fig. 3). It is clear from this figure that
there is a considerable phase lag between the months with
high local rainfall and the month of highest river level.
Water level at this location reflects the total of rain-
water run-off from all the upriver regions, added to by
the local precipitation run-off. The amplitude peak in
June is partially due to the greatly increased water volume from the upper Rio Negro and Rio Branco, both areas
having the greatest amount of rain during May to July.
It must be understood that an increased water influx in
the headwaters will be reflected with very little delay,

perhaps a day, in a level increase 1000 km downriver, while it takes about 15 days for the local run-off water to reach the same downstream point, assuming 1.0 to 1.5 m/ sec velocity.

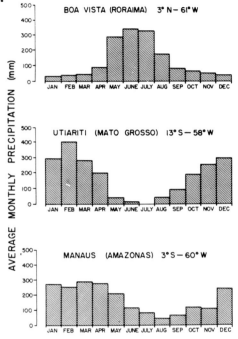

Fig. 2. Average rainfall data for three selected stations exemplifying the three prevailing climate types of the Amazon basin: Boa Vista representing the northern tropical savanna climate with major rains falling April to August; Utiariti on the Brazilian plateau as an example for the southern savanna climate where rainy and dry seasons are occurring at a time of year opposite to the northern savanna climate; and Manaus showing the typical rain forest climate where a really dry season is lacking. (Data taken from: Galvão, 1959).

A similar phenomenon can be observed in the lowest reaches of the Amazon where the ocean tides cause significant water level changes several hundred km upstream although the current reversal due to the tides is restricted to a smaller distance (Egler and Schwassmann, 1962). Tidal water level changes of 1.5 to almost 4.0 m are observed in the Baía de Marajó, the smaller southern arm of the Amazon delta.

In the upper and middle regions the annual precipitation differences cause considerable water level changes correlated with a greater water volume flowing downriver and higher current velocity. In the estuary no such seasonal

water level differences are noted but the different river
discharge rates cause a horizontal displacement of the
mixing zone of river with ocean water of about 200 km.
    A consequence of the water level increase is the flood-
ing of vast adjacent areas, the floodplain lakes. These
varzea lakes, especially noticeable in the silt-laden
Solimões and lower Amazon, contain only little water during
the low water season and are separated from the main river
bed by extensive low banks, the result of sedimentation
of these white waters while clear run-off water from local
rains enters the varzea peripherally.

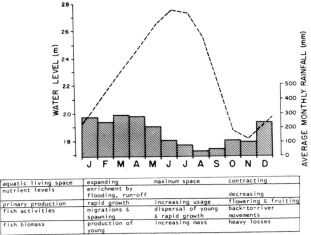

aquatic living space	expanding	maximum space	contracting
nutrient levels	enrichment by flooding, run-off		decreasing
primary production	rapid growth	increasing usage	flowering & fruiting
fish activities	migrations & spawning	dispersal of young & rapid growth	back-to-river movements
fish biomass	production of young	increasing mass	heavy losses

Fig. 3. Annual cycle of events in the varzea lakes. Rainfall data and
river level changes (broken line) at Manaus harbor. Water level in the
floodplain lakes is determined by the river water level. (after an
idea by Lowe-McConnell, 1975).

    Conditions in the Rio Negro are somewhat different from
those in the Solimões. Lack of sediment apparently is the
reason for the absence of typical varzea lakes that become
physically separated from the main river. Flooding of bord-
ering forest areas and lower reaches of small tributaries
causes the typical 'igapó' biotope. It can be assumed that
conditions similar to those in varzea lakes are present
also in the igapó habitat, e.g. higher nutrient levels and
an increase in space. The influx of some white water rivers
into the Rio Negro and their appreciable sediment load is
probably the cause for the enormous area of low islands
(Anavilhanas) that are under water during high river level
and that divide the main river channel into an anastomosing
network of 'furos'.

Spawning Runs and Reproduction in Floodplain Lake Species
Considerable information exists concerning the seasonal

movements, spawning, and occurrence of young, for the
Rupununi savanna of Guyana, the Venezuelan savannas, and
the Paraná-Paraguai drainage and has been reviewed by Lowe-
McConnell (1975). A few observations have been made in the
Janauacá varzea lakes of the lower Solimões near Manaus.
Most of the characoid species that are found in this lake
system are total spawners, all the eggs ripening and being
shed at one time. Some of the important food fishes in
this group are the Tambaquí (*Colossoma bidens*), Curimatã
and Jaraquí (*Prochilodus* sp.), Branquínha (*Anodus* sp.),
Matrinchã (*Brycon* sp.), and Peito de aço (*Potamorhina
pristigaster*). Females of these species have fully developed ovaries in January-February, the beginning of the
rainy season when water levels in the lake system start
to rise. In the curimatã the ovary is approximately one
fourth of the total weight of the female and can contain
up to 300,000 eggs, each of almost 1.0 mm diameter. These
fish begin their characteristic short-distance migration
out of the lakes into the many furos that connect to the
main river at the time of rising water levels and at the
time specific for each species. If these migrations occur
in schools they are called 'Piracema'. No evidence seems
to exist regarding the actual locale of spawning, but the
fish return to the lake with spawned-out ovaries. It is
believed that Curimatã spawn in flooded shore areas of the
furos.
    A diagram which summarizes conditions during the annual
cycle in these varzea lakes is shown in Fig. 3.
    The adaptive significance of the coincidence of spawning
and rising water levels seems to be in the increasing living space and food supply which become available at this
time. Stomach analysis of fish species in varzea lakes
(Marlier, 1968; Knöppel, 1970), even though they were restricted to a particular season only, indicate that most
species are capable of utilizing different kinds of foods
rather than being specialized stenophages. A recent three-
year study could demonstrate the food items quantitatively
on a month-to-month basis in the tambaquí (Honda, 1974).
These locally important food fishes turn out to be omnivor-
ous, being specialized for feeding on several different
food items. A well developed gill raker apparatus enables
them to filter-feed on planktonic crustacea predominantly
during the dry season while their strong crushing teeth
permit feeding on hard fruits during the high water season.
    In addition to the already listed commercially important
species which appear so well adapted as total spawners to
the suddenly-opening opportunity of increased living space
and abundant food supply, there are other species of fish
of different suborders that are partial spawners. Some of
these, as the many species of cichlids, produce successive
broods and show parental care. Of commercial value is the
pirarucú (*Arapaima gigas*), the largest freshwater fish

known. On the other end of the size scale are many species
of gymnotoids most of which are probably also partial
spawners. Cichlids and gymnotoids are known to remain permanently in these floodplain lakes and are often found to
enter into the many small igarapés that flow into these
lakes from the higher ground of the terra firma. Most of
these small fish feed on aquatic insect larvae which are
present almost year round but never very abundant. Producing fewer young in several successive spawning bouts over
a long period could also be considered of adaptive value
for these species. Many of these small fish also occur in
the island 'archipelago' of the Rio Negro and although
being partial spawners which reproduce over several months,
preliminary sampling data of populations of the same species from both locations indicate synchronous spawning activity.

Small Forest Streams, Igarapés

Rainfall, run-off and floods   These many small and medium
sized streams of the terra firma, the higher not inundated
ground, make up the other predominant type of aquatic habitat. Because of their great numbers, their total surface
areas combined, as well as their total lengths, are probably more than a thousand times that of the main rivers
including the floodplain lakes, even though each one of
these igarapés by itself may look rather insignificant.
These streams are spring-fed and permanent, run-off from
local rains is added frequently. There are those with more
extensive drainage areas that show a somewhat greater volume and elevated water level during the season with increased precipitation, but there are many that exhibit
almost no appreciable seasonal differences in water level.
This aquatic environment is the opposite extreme to the
floodplain lakes because annual differences in water volume and water quality are barely noticeable. A small but
recognizable annual change consists of more frequent short-
term flooding due to local rains in the wet season. As far
as I could determine, all teleosts living in this habitat
are partial spawners with several successive egg-laying
sessions. These are the gymnotoids, the cichlids, some
catfish, and some characoid species as the traíra (*Hoplias*),
Astyanax, and others.
    Being of low nutrient content and receiving little solar
radiation, these igarapés are of low primary productivity
and the surprisingly rich and diverse ichthyofauna makes
good use of all kinds of foods including considerable
allochthonous material, mostly small invertebrates and
plant parts (Knöppel, 1970).

Reproductive Timing of Igarapé Species   One of the highly
specialized gymnotoid electric fish, the sandfish *Gymnorhamphichthys hypostomus*, has been studied since 1964 in
several parts of the Amazon and adjacent basins (Schwass-

mann, 1976). The principal habitat of this nocturnal electric fish are small streams with clean sand deposits on
the bottom in which these fish burrow during the day.
Occasional specimens are found in larger rivers, but these
may be stragglers.

As probably all gymnotoids, *Gymnorhamphichthys* is a
partial spawner and reproductive activity seems to consist
of four successive spawning bouts indicated by the presence of four, or less, size classes of eggs in the female's
ovaries. Earlier data seemed to indicate a two-year life
span and a restriction of reproductive activity to a few
months of the year. The presence of two distinct size
classes, mature spawners and smaller immatures, suggested
that the fish might begin spawning activity only towards
the end of the second year and probably die after complet-
ing reproduction (Schwassmann, 1976). Continued field work
during the last two years, especially repeated sampling
of populations at collecting sites of previous years and
additional new population estimates, seems to permit a
re-evaluation of the earlier data. The four diagrams
(Fig. 4 a-d) show the composition of samples collected
near Belém, Manaus, in Rondônia, and in Colombia. Size
and gonadal status of the fish are plotted at the time of
year when they were collected. The bar graph on the bottom
indicates average monthly precipitation in the general
area of the sampling sites.

Without going into details, present interpretation of
the combined data is that spawning activity must be going
on for at least six months of the calendar year. The presence of a substantial class of immatures that are defin-
itely much larger than the newly-born young of current
spawning bouts, could indicate two major peaks of reproductive activity per year which would also imply that the
fish mature within one year. Synchronization of gonadal
ripening and spawning activity is evident within each population, but populations of nearby streams may be at a
different developmental stage. Those populations which
live in regions exhibiting a more distinct separation of
the year into wet and drier season, or live at 10°S latitude (Rondônia) where there exists an annual daylength
change of about one hour, indicate in their size composition and maturity status better synchronization than
populations at Manaus and Belém where seasonal environmental factors fluctuate less. Young sandfish only about
one to two weeks old were found and peculiar behaviour
patterns of mature fish were observed that are interpreted
as spawning activity. Each sampled population seems a unit
in itself and intrapopulation synchronization of reproductive events must be the prevailing mechanism assuring
successful spawning. Deposition of eggs and their fertil-
ization takes place at the very site where the population
lives and no migratory movements seem to occur.

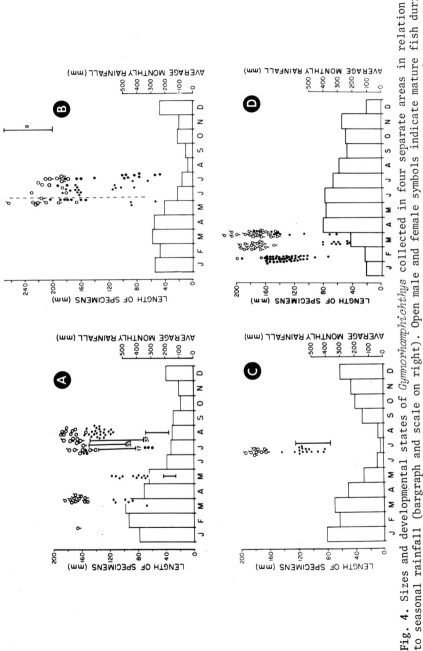

Fig. 4. Sizes and developmental states of *Gymnorhamphichthys* collected in four separate areas in relation to seasonal rainfall (bargraph and scale on right). Open male and female symbols indicate mature fish during spawning, dots represent immature specimens; vertical bars (and numbers) stand for the range of a large sample of immatures (in *a* and *c*), or mature fish (in *b*). Collecting sites: Belém (E in Fig. 1) in *a*; Manaus (F in Fig. 1) in *b*; Rondônia (B in Fig. 1) in *c*; Colombian llanos (G and H in Fig. 1) in *d*. Rainfall data for *a*, *b*, *c*, from Galvão, 1959; those of *d* are a composite from various sources.

The Island of Marajó, a Unique Situation

Rainfall and Floods   The interior of this largest estuarine island (40,000 km^2) undergoes drastic seasonal changes that affect the local fish fauna and the important fisheries drastically. Some early work was done in 1960-61 to document these profound seasonal events (Egler and Schwassmann, 1962) but no further and more extensive study has come to my attention. In contrast to the central Amazon basin and also to the city of Belém, rainy and dry season are very distinct on Marajó. January to May are the months with high precipitation while only little rain falls the remaining months. The vast amount of rain water drains only very slowly through a few rivers and most of it forms a huge lake on the eastern half of the island which is not covered by rainforest. During the rainy season only a few elevated plateaux provide refuge for cattle and other animals, while the island is covered by about 2 m of water. Considerable evaporation during the dry months causes a high concentration of salts, the water resembles a salty brine and becomes unpotable. These concentrated waters become subsequently confined to a few shallow lakes that contain a high density of marketable and smaller fish.

Reproductive Timing   Conditions on Marajó resemble very closely those in the Rupununi savanna and the reproductive activities of existing fish species show similar synchronization with the environmental cycle. Spawning of many species was observed to occur on recently flooded areas after the onset of heavy rains in February. Marajó has one of the richest fish faunas in the Amazon, several representatives of marine families included, and seems to be of importance in understanding the evolution and the pathways of spreading of newly developing species of primary freshwater fishes.

Experimental Studies Concerning Annual Reproduction and Possible Timing Mechanisms in Amazonian Fishes

Earlier work and suggestive evidence related to possible environmental factors in the timing of gonadal maturation has been reviewed recently (Schwassmann, 1971). Almost nothing is known about fish from central Amazonia, but some of the early experimental work at the fishculture stations Lima Campos and Fortaleza, Ceará, Brazil, must be mentioned, since most of their original stock of fish came from the Amazon through their early collaboration with the Museu Emilio Goeldi at Belém, Pará.

The method of pituitary treatment for controlled inducement of spawning was developed and applied to pond culture in Ceará since 1930. Fontenele *et al.* (1946) could obtain three successive spawnings by injection of pituitary extracts in *Prochilodus* which is a total spawner. What environmental factors are involved in any possible long-

term regulation of gonadal ripening and of actual release
of migratory and spawning behavior remains a mystery. In
the Amazonian gymnotoid *Eigenmannia* successful gonadal
maturation, spawning and rearing of young was accomplished
in aquaria by Kirschbaum (1975). Unfortunately, a combination of simulated 'normal' factors was employed, as water
level increase, lowering of the pH, rain simulation in
addition to the non-natural 13 h light - 11 h dark lighting regimen. Other equatorial vertebrates were found to
respond to photoperiodic induction with gonadal ripening
(Miller, 1959), and a possibly similar effect, but of
decreasing day-length, was suspected for fish populations
at 10^0 S latitude (Schwassmann, 1976). If it could be
shown that commercially valuable food fishes of central
Amazonia responded to photoperiodic induction, as the
study on *Eigenmannia* could be interpreted to indicate,
an easy way for controlling spawning activities in future
pond cultures would become available.

Concluding Remarks

Available evidence supports the view that those fishes
which live in central Amazonian floodplain lakes and which
are total, one-time/year, spawners begin migrations and
subsequent spawning during the time of rising water level.
Spawning probably occurs in shallow recently flooded areas.
The many species of characoid, cichlid, gymnotoid, and
nematognath fishes inhabiting small streams in the forests
at higher elevations are probably all partial spawners
that go through several successive spawning sessions.
Synchronization of the members within one population appears precise but different populations in adjacent
streams can be at different developmental stages. Since
this particular habitat, the igarapés of the firm ground,
shows only minimal annual changes in water level, due to
rain-water run-off, and in water chemistry, it is here
that one would have to look for possible free-running,
circ-annual rhythms, a possibility mentioned by Bünning
(1967). The most extensive data so far have been collected
on the small sand-dwelling electric gymnotoid *Gymnorhamphi-*
chthys and should provide a basis for further experimental
studies.

References

Bünning, E. (1967). *The Physiological Clock*. 2nd rev. ed. Springer,
    Berlin, pp.167.
Egler, W.A. and Schwassmann, H.O. (1962). *Bol. Mus. Geoldi N. S.
    Avulsa* 1, 2-25.
Fontenele, O., Camacho, E.C. and de Menezes, R.S. (1946). *Bol. Museu
    Nacl. (Rio de Janeiro), Zool.* 53, 1-9.
Galvão, M. Velloso (1959). *In: Geografia do Brazil - Grande Região
    Norte, I.B.G.E.*, Conselho Nacional de Geografia, Rio de Janeiro,
    pp. 61-111.

Géry, J. (1969). *In: Biogeography and Ecology in S. America,* Fittkau
  et al., (ed.), The Hague, **2**, 828-848.
Honda, E.M.S. (1974). *Acta Amazonica* (IV) pp. 47-53.
Kirschbaum, F. (1975). *Experientia* **31**, 1159-1160.
Knöppel, H.A. (1970). *Amazoniana* **2**, 257-352.
Lowe-McConnell, R.H. (1964). *J. Linn. Soc. (Zool.)* **45**, 103-144.
Lowe-McConnell, R.H. (1969). *Biol. J. Linn. Soc.,* **1**, 51-75.
Lowe-McConnell, R.H. (1975). *Fish Communities in Tropical Freshwaters.*
  Longman, London, pp. 337.
Marlier, G. (1968). *Cadernos Amazonia,* **11**, 21-57, Instituto Nacional
  de Pesquias da Amazonia, Manaus, Brazil.
Miller, A.H. (1959). *Condor* **61**, 344-347.
Nijssen, H., Isbrücker, I.J.H. and Géry, J. (1976). *Stud. Neotropic.
  Fauna* **11**, 37-63.
Oltman, R.E., H. O'R. Sternberg, F.C. Ames and L.C. Davis, Jr. (1964).
  Geol. Surv. Circular, Washington. No. 486.
Roberts, T.R. (1972). *Bull. Mus. Comp. Zool. Harv.* **143**, 117-147.
Schwassmann, H.O. (1976). *Biotropica,* **8**, 25-40.
Schwassmann, H.O. (1971). *Fish Physiology* **6**, 371-428, Academic Press,
  New York.
Sioli, H. (1965). *Amazoniana* **1**, 74-83.
Soares, L. de Castro. (1959). *In: Geografia do Brasil,* **1**, Rio de
  Janeiro. pp. 128-194.

LUNAR AND TIDAL RHYTHMS IN FISH

R.N. GIBSON

*Dunstaffnage Marine Research Laboratory,
Oban, Argyll, Scotland*

Introduction

The study of rhythmic changes in the behaviour and physiology of living organisms has concentrated for the most part upon rhythms which are related to the solar day. There are, however, many instances in which the activities of animals and plants are synchronised with environmental cycles which owe more to the influence of the moon than to that of the sun. The moon's influence is most marked in the sea mainly because of its effect upon the tides and consequently the majority of examples of rhythms with lunar or tidal periods are found in marine animals. It is possible to distinguish between three types of lunar-related rhythms; those whose periods approximate to a synodic lunar month of 29.5 days (lunar rhythms), to half a lunar month of 14.7 days or the interval between two consecutive sets of spring tides (semi-lunar rhythms) or to a tidal cycle of 12.4 hours (tidal rhythms). Tidal rhythms may also be considered as bimodal lunar day (24.8 hours) rhythms (Palmer, 1974).

Reviews of the literature dealing with lunar and tidal rhythmicity are given by Korringa (1947, 1957), Fingerman (1957, 1960), Hauenschild (1960), Palmer (1974) and Enright (1975) who deal predominantly with invertebrates, although McDowall (1969) and Schwassmann (1971) include many fish among their examples. The aim of this paper is to provide a comprehensive and up-dated account of published work dealing with lunar and tidal rhythms in fish.

Lunar and Semi-Lunar Rhythms

Lunar rhythms, in which a single peak of activity occurs each month, seem to be rare in fish. The best documented example is that of the European eel (*Anguilla anguilla*) in its migration to and from the sea. At the time of their seaward migration, although some individuals

may be found migrating down the rivers at all times of
the lunar month, by far the greatest proportion is caught
in eel traps when the moon is waning. Such movements are
at a minimum on the full moon (Fig. 1a,b,c).

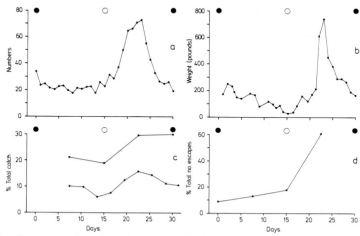

Fig. 1. Lunar rhythm in the seaward migration of eels. (*a*) In the
upper Rhine (from data in Jens, 1953); (*b*) In the Baltic (redrawn
from Fig. 4 in Jens, 1953); (*c*) In Holland. Upper curve for the
Ijsselmeer and lower curve for the Dutch canals (from data in Deelder,
1954); (*d*) % of total escapes of eels from experimental containers
related to the lunar month (from data in Boëtius, 1967). The time of
the new moon is shown by the closed circles and the full moon by open
circles.

At first sight the correlation between the phase of the
moon and the extent of migration suggests that the intensity of moonlight is the controlling factor. Cloud cover,
however, makes little difference to the overall pattern
and in the course of experiments on the times of escape of
eels from light-proof containers, most attempted to escape
when the moon was in its last quarter (Boëtius, 1967)
(Fig. 1d). These results imply that some form of endogenous control is involved in the eel's seaward migration.
In certain areas the elvers also show a periodicity in
their movements. When migrating up the River Elde from
the sea the majority are caught at the times of the last
quarter and new moon (Gollub, 1959) (Fig. 2), although
Meyer and Kuhl (1953) found little correlation between
moon phase and elver migration in the River Ems. The relationship between eel migration and lunar factors has
been summarised by Deelder (1970) and Tesch (1973).
    A curious, and as yet unexplained, phenomenon, is the
lunar periodicity in the response of the guppy (*Lebistes*

reticulatus) to coloured light. When illuminated from
above the fish maintains an upright position with its
dorso-ventral axis vertical, but when the light is shone
from the side the fish leans towards it (the dorsal light
response).

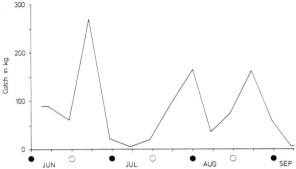

Fig. 2. Lunar rhythm in the catches of ascending elvers in the River
Elde (redrawn from Fig. 3 in Gollub, 1959) Symbols as Fig. 1.

Its angle of inclination lies somewhere between the vertical and the angle of incidence of the light source. Using
this angle as a measure of sensitivity to different wave
lengths, Lang (1965, 1967, 1969, 1970) found that the
fish was most sensitive to yellow light (583 nm) over the
full moon period and least during the new moon. The reverse
was true for red (670 nm) and violet (423 nm) light, but
responses to other parts of the spectrum did not fluctuate
markedly.
    There are several accounts of variations in the catch
of commercial species which correlate with the moon's
phase, most of which refer to clupeids. British catches
of the Atlantic herring (*Clupea harengus*) in the North
Sea showed a peak at full moon for several pre-war years.
The exact timing of this peak varied from year to year,
however, and good catches were only made when the full
moon occurred between the middle of October and early
November (Fig. 3) (Savage and Hodgson, 1934). In the German
post-war catches there were unequal peaks just after the
full and new moons (Jens, 1953, 1954). Catches of hake
(*Merluccius merluccius*) were also greater at full moon in
the pre-war years (Hickling, 1927) but the catches of
Pagellus centrodontus were higher at the new moon (Desbrosses, 1932). There are similar reports of lunar periodicity in the catch of *Clupea pallasii* (Tester, 1938) and
the Californian sardine (Clark, 1956). Blaxter and Holliday
(1963) reviewing the evidence for lunar cycles in clupeid
catches, put forward several possible explanations involving fluctuations in the intensity of moonlight and effects
of the tide, but also suggest that the results may be in-

dicative of the behaviour of the fishermen rather than
the fish.

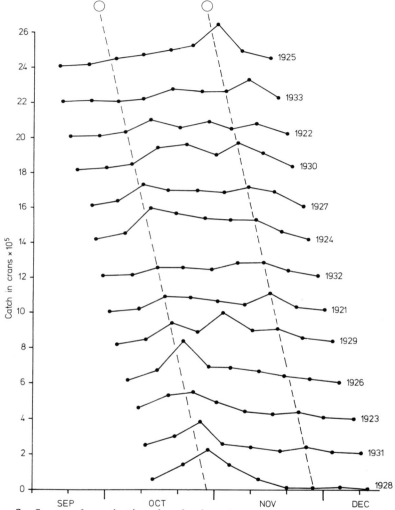

Fig. 3. Seasonal variation in the herring catches in the North Sea
from 1921-1933. Each graduation in the vertical axis represents
2×10^5 crans and the dashed lines join the times of the full moon
(from data in Savage and Hodgson, 1934).

The spawning behaviour of several species is adapted
to take advantage of the bi-monthly variation in tidal
amplitude associated with the spring-neap cycle, and at
high water of spring tides lay their eggs high on the

shore where they are presumably less vulnerable to aquatic predators. The most famous example of such a semi-lunar spawning rhythm is that of the California grunion, *Leuresthes tenuis*. The breeding season of this fish lasts from late February to early September but actual spawning only takes place during the night on the three to four high tides of descending amplitude following each new or full moon (Fig. 4a).

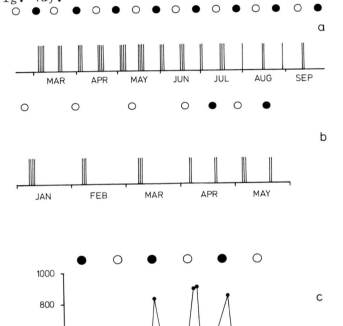

Fig. 4. Semi-lunar spawning rhythms. (*a*) In *Leuresthes tenuis* showing the relationship of the spawning runs in 1946 (vertical lines) with the phase of the moon. (adapted from Fig. 6 in Walker, 1952); (*b*) In *Galaxias attenuatus*. The symbols are the same as (*a*) above (from data in Hefford, 1931); (*c*) In *Enchelyopus cimbrius*. Density of spawned eggs collected in plankton tows at different times of the lunar month (redrawn from Fig. 2 in Battle, 1930).

The timing of the spawning runs is apparently not caused directly by the tides but by the interaction of some undis-

covered lunar factor with an approximately 18 day rhythm
of gonad maturation. The eggs are laid and fertilised in
the top few centimeters of wet sand and continue to develop
there until the next series of spring tides. The fully
developed larvae hatch rapidly under the agitating influence of the waves and are swept out to sea, but if the
eggs are not covered by water they are able to survive for
another fortnight until the next set of high tides (Thomson, 1919; Clark, 1925; Walker, 1949, 1952). The reproductive behaviour of the Gulf of California grunion, *Leuresthes
sardina*, is very similar in outline to that of *L. tenuis*
except that it spawns during the day as well as the night
(Walker, 1952), and its spawning runs seem to be controlled
by a response to tide height rather than to moon phase as
such (Thomson and Muench, 1976). Analyses of the timing
of the spawning runs suggested that daytime spawning is
related to the fact that there is a shift of the time of
the higher spring tides from the early morning to the late
afternoon in the middle of the spawning season. There is
also an important difference in the duration of the spawning act between the two species. *L. tenuis* takes about 30
seconds to lay and fertilise its eggs while *L. sardina*
usually takes only 3-4 seconds to complete this process
(Thomson and Muench, 1976). Thomson and Muench (1976) consider that the difference evolved not because of increased
predation pressure by birds on the daytime spawning Gulf
grunion, but as a response to the short period waves in
the Gulf. On the open Pacific coast the wave periods are
longer and allow *L tenuis* more time to complete its spawning before being disturbed by the next wave. The surf
smelt (*Hypomesus pretiosus*) has comparable spawning habits
in that the eggs are laid on the beach at a time when the
highest high tides occur during the day. In this species
the eggs are not buried by the females but are washed into
the sand by wave action (Thomson, 1936; Loosanoff, 1938).
The four-eyed fish *Anableps microlepis* from the Amazon
estuary also takes advantage of high spring tides in March
to deposit its young in brackish water lagoons not flooded
during neap tides (Schwassmann, 1967).
     The spawning cycle of the New Zealand fish *Galaxias
attenuatus* is of particular interest in the context of
semi-lunar rhythms. Although essentially a freshwater
species, the larvae are marine. During the spawning season, from January to May, the mature adults migrate down
the rivers and lay their eggs among the vegetation on the
banks of estuaries. The timing of this migration has a
well-defined lunar or semi-lunar periodicity although the
exact phase relationship varies from year to year (Burnet,
1965). Hefford (1931) gives details of the spawning times
of the same species in the Manawatu river where there was
a clear correlation between the date of spawning and the
full and new moons (Fig. 4b). Once laid, the eggs remain

and develop in the grass, hatching on the next series of spring tides.

The synchronisation of the spawning of *Leuresthes*, *Hypomesus* and *Galaxias* with spring tides clearly has some adaptive advantage to the species, but in the case of the offshore gadoid, *Enchelyopus cimbrius*, which also spawns predominantly on spring tides, the advantage is less obvious. In a series of plankton hauls off the eastern Canadian coast throughout the spawning season, the greatest number of eggs was collected at the times of the new and full moon (Battle, 1930) (Fig. 4c).

How the periodic spawning of the grunion and other species with comparable habits is controlled has yet to be fully explained, although, because the gonads begin to develop in advance of the actual spawning runs, some form of endogenous control is probably involved. The factors regulating the lunar periodicity in the behaviour of the commercial species mentioned above, if it is a real phenomenon, are likewise unknown.

Tidal Rhythms

In addition to those fish which exhibit rhythms of lunar or semi-lunar frequency in their behaviour, there are a large number of species which synchronise their activities with the tidal cycle. Most of such fish inhabit the intertidal or immediately sub-tidal zones and consequently are subject to continuous changes in their environment. The literature covering this subject prior to 1969 has been reviewed in an earlier paper (Gibson, 1969a). Since that time, numerous additional examples of tide-related movements have been described for species inhabiting the whole range of intertidal habitats. Such fish may be broadly classified as permanently inhabiting the littoral zone for the majority of their life, living there in their juvenile stages or only visiting it at high tide. These three categories have been termed true residents, partial residents and tidal visitors (Gibson, 1969a) or primary residents, secondary residents and transients (Thomson and Lehner, 1976). On rocky shores the true or primary residents belong mainly to the families Blenniidae, Gobiidae and Gobiesocidae, together with some members of the Cottidae and are well adapted to the rigors of intertidal conditions. For the most part they are least active at low tide and remain quiescent in rock pools, beneath stones, or in crevices. Over the high tide period they forage together with the tidal visitors in regions which are exposed at low water (Zander, 1967; references in Gibson 1969a; Green 1971; Thomson and Lehner, 1976). In other types of intertidal habitat such as sandy and muddy shores or tidal creeks, where there is little cover available at low water, the fish migrate in with the flood tide and leave on the ebb. The exceptions are the amphibious mudskippers (Perio-

phthalmidae) which feed and defend territories on the mud-
flats of mangrove swamps when the tide is out (Brillet,
1975). Cain and Dean (1976) investigated annual cycles of
diversity and abundance of the fish moving into tidal
creeks and considerable attention has been focussed on
the intertidal movements of the juveniles of commercial
flatfish on sandy beaches (Tyler, 1971; Gibson, 1973a;
Kuipers, 1973). These fish move into the intertidal zone
on a rising tide, feed, and move out on the falling tide.
The exact pattern of feeding varies widely in different
localities. Plaice (*Pleuronectes platessa*) migrating on
to the extensive sand flats of the Dutch Wadden Sea feed
only on these flats and not in the channels to which they
retreat at low tide (Kuipers, 1973), whereas on the open
coast they feed over the whole tidal cycle, but most intensively shortly after high tide (Thijssen *et al.*, 1974).
In other areas this species feeds continuously over the
whole tidal cycle without any definite tidal periodicity
(Edwards and Steele, 1968; Gibson, 1973a). The goby, *Poma-
toschistus (Gobius) minutus* frequently found in the same
habitat and exhibiting similar migratory behaviour to the
plaice, has a marked tidal periodicity in its food intake.
The juveniles feed most intensively at high tide whereas
the guts of the adults are fullest on the ebbing tide
(Healey, 1971).

Laboratory experiments on the behaviour of these intertidal fish under controlled conditions have demonstrated
the presence of persistent rhythms of approximately tidal
(circatidal) frequency. Gompel (1937) claimed that plaice
show a tidal rhythm of oxygen uptake in the laboratory
and Day (1968) suggested that *Fundulus similis* is most
resistant to the toxic effects of endrin and sodium chlo-
ride at high tide. Similarly, Schwarz and Robinson (1963)
postulated that there may be a correlation between the
rate of oxygen uptake of *Ospanus tau* and the phase of the
tidal cycle. The majority of laboratory studies have, however, been concerned with rhythms of locomotor activity.
Among the Blenniidae, persistent circatidal rhythms have
been described in *Blennius pholis* (Gibson, 1965, 1967)
(Fig. 5a), *Coryphoblennius galerita* (Gibson, 1970) (Fig.
5b) and *Blennius cristatus* (Stahl, 1973). All three species
exhibit well marked rhythms in LL or DD with peaks of activity synchronised with the predicted time of high water.
In LD cycles the tidal peaks of both *B. pholis* and *C.
galerita* are depressed during the dark period (Fig. 5c,d),
whereas darkness enhances activity in *B. cristatus*. The
clingfish *Tomicodon humeralis* from the Gulf of California
also possesses a persistent tidally-phased activity rhythm,
but in this species both the presence of food and darkness
suppress activity. The rhythm of another clingfish, *Gobiesox pinniger*, living in the same habitat but at a lower
level on the shore, persists for only two cycles (Gollub,

1974). Similar short-lived rhythms have also been described
in the tide-pool cottids *Oligocottus maculosus* from the
Pacific coast of Canada (Green, 1971) and *Acanthocottus
bubalis* on the British coast (Gibson, 1967). The tidal
rhythms of juvenile plaice, flounder (*Platichthys flesus*)
and turbot (*Scophthalmus rhombus*) also persist for only
2-3 cycles in the laboratory and rapidly revert to a bimodal circadian pattern (Gibson, 1973b, 1976).

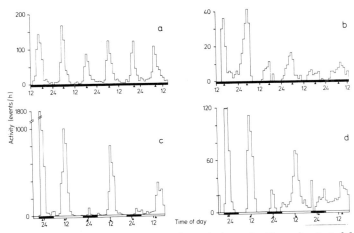

Fig. 5. Examples of tidally-phased activity rhythms in two blennies
(*a*) *Blennius pholis* in DD (from Gibson, 1965); (*b*) *Coryphoblennius
galerita* in DD (from Gibson, 1970); (*c*) *Blennius pholis* in LD (from
Gibson, 1971); (*d*) *Coryphoblennius galerita* in LD (from Gibson, 1970).
Predicted high tides are shown by black triangles.

The rhythm of the estuarine flatfish, *Trinectes maculatus*,
however, shows recognisable peaks for at least six cycles
in continuous illumination. In LD those tidal peaks occurring during the light period are suppressed, resulting in
a unimodal rhythm (O'Connor, 1972).
     In contrast, the circatidal rhythms of the Japanese
mudskippers *Boleophthalmus chinensis* and *Periophthalmus
cantonensis* may persist for up to 50 days in constant conditions (Ishibashi, 1973; Nishikawa and Ishibashi, 1975).
Boleophthalmus also shows a semi-lunar rhythm in the height
of its activity peaks which Ishibashi (1973) considered
to be the result of the interaction between circadian and
circatidal components.
     Comparison of the same or related species from tidal
and non-tidal conditions has shown that their rhythms have
a character which reflects the relative importance of the
tidal regime in their particular environment. Young floun-
ders (*Platichthys flesus*), for example, inhabiting beaches
with marked tidal fluctuations have a weakly persistent

circatidal rhythm which soon changes to a bimodal circadian
rhythm (Gibson, 1976). The same species in the Baltic,
where tidal ranges are very small, shows only a bimodal,
or possibly trimodal, circadian rhythm in the laboratory
(Muus, 1967). Similarly two species of *Blennius* from the
Mediterranean are most active at dawn and dusk (Gibson,
1969b) but the rhythm of related species on the British
coast is phased with the tides (Gibson, 1967, 1970). In
this context, Naylor (1976) has pointed out that the
rhythms of species which are found in both tidal and non-
tidal environments are usually predominantly circadian
and any tidal component that may be present is rapidly
lost in non-tidal conditions.

The factors responsible for the entrainment of tidal
rhythmicity in fish are less well known than those which
induce and phase the rhythms of invertebrates, and crustacea in particular. Periodic agitation by water and sand
particles is the most effective zeitgeber for the tidal
rhythm of the isopod *Excirolana chiltoni* (Enright, 1965)
but several environmental stimuli are involved in the
entrainment of the tidal rhythm of the crab *Carcinus
maenas*, notably cyclic changes in pressure, temperature
(Naylor *et al.*, 1971) and salinity (Taylor and Naylor,
1977). In the intertidal fish *Blennius pholis* hydrostatic
pressure cycles are capable of entraining and resetting
the phase of the rhythm (Gibson, 1971). Pressure cycles
are probably also involved in entraining tidal rhythmicity
in *Pleuronectes platessa* (Gibson, *et. al.*, 1978) and circ-
umstantial evidence suggests that pressure may be impli-
cated in the synchronisation of the rhythm of *Oligocottus
maculosus* (Green, 1971). The activity rhythm of *Boleophthalmus chinensis*, however, can be entrained in fish which
have become arhythmic in the laboratory by exposing them
to a single change in water level (Ishibashi, 1973) and
in *Periophthalmus cantonensis*, it is possible to entrain
a 12 or 24 hour rhythm by feeding the fish at 12 or 24
hour intervals (Nishikawa and Ishibashi, 1975).

At this point it is appropriate to consider how the
activity rhythms recorded in the laboratory compare with
the overall pattern of behaviour of the fish in their
natural environment. In the few species where such comparisons have been made considerable differences are
apparent.

Oligocottus maculosus shows a weakly persistent (24-48
hours) tidal rhythm in the laboratory. In this rhythm the
activity period lasts for 2-3 hours and its mid-point
corresponds to the time of the concurrent high tide. In
the field, activity is regulated by temperature, turbulence
and light intensity and under certain combinations of these
factors, the fish may be totally inactive at high tide
(Green, 1971). *Blennius cristatus* also exhibits tidally-
phased rhythmicity in the laboratory. In LD the tidal

peaks expected during the light period are suppressed
and from such a pattern it might have been inferred that,
in the sea, the fish would be active during nocturnal high
tides. Field observation of this species in its natural
habitat, a fossilised mangrove reef in Florida, demonstrated, however, that the reverse is true; it is most
active at low tide during the day (Stahl, 1973). This
puzzling difference between the fishes behaviour in the
laboratory and the field still remains mostly unexplained.
A third, although not as extreme, discrepancy between
field and laboratory results is found in juvenile plaice.
In the laboratory the fish are inactive at the expected
time of low water and early flood and most active on the
early ebb. In the sea they are active over the whole tidal
cycle (Gibson, 1975).

Such discrepancies only serve to emphasise the dangers
inherent in assuming that because a species exhibits a
particular activity pattern in the laboratory, the nature
and timing of its behaviour will be similar in the wild.
They also suggest two further questions. First, what contribution, if any, does the endogenous rhythm make to the
activity pattern observed in nature, and secondly, what
does the activity measured in the laboratory represent?

In the case of *Oligocottus maculosus*, the activity recorded in the laboratory is considered to be a manifestation of a 'biological clock' which, coupled with an escape
response, enables a fish displaced from its home pool to
return at the time of high tide (Green, 1971). The rhythmic activity of juvenile plaice in constant conditions,
on the other hand, possibly represents an internally controlled change of an innate tendency to move into deeper
water. In the laboratory the fluctuating intensity of
this behaviour is expressed as a rhythmic change in activity, reaching a maximum at the expected time of the ebb
tide. In the sea the role of the rhythm may be to ensure
that the fishes downshore migration is correctly phased
(Gibson, 1975).

The results of these comparisons strongly suggest that
in any studies of rhythmic behaviour designed to elucidate
ecological problems, the quality as well as the quantity
of activity being measured should be examined. Only in
this way can meaningful ecological inferences be drawn
from laboratory studies of rhythmic activity patterns.

References

Battle, H.I. (1930). *Contrib. Can. Biol. Fish.*, N.S. **5**, 363-380.
Blaxter, J.H.S. and Holliday, F.G.T. (1963). *In: Advances in Marine
    Biology* (F.S. Russell, ed.), **1**, 261-393. Academic Press, London.
Boëtius, J. (1967). *Meddr Danm. Fisk.-og Havunders*, **6**, 1-6.
Brillet, C. (1975). *Z. Tierpsychol.*, **39**, 283-331.
Burnet, A.M.R. (1965). *N.Z. Jl. Sci.*, **8**, 79-87.
Cain, R.L. and Dean, J.M. (1976). *Mar. Biol.*, **36**, 369-379.

Clark, F.N. (1925). *Calif. Div. Fish Game, Fish. Bull.*, **10**, 1-51.
Clark, F.N. (1956). *Calif. Fish Game*, **42**, 309-323.
Day, J.W. (1968). *Proc. La Acad. Sci.*, **31**, 62-64.
Deelder, C.R. (1954). *J. Cons. perm. int. Explor. Mer*, **20**, 117-185.
Desbrosses, P. (1932). *Revue Trav. Off. Pech. marit.*, **5**, 167-222.
Edwards, R. and Steele, J.H. (1968). *J. exp. mar. Biol. Ecol.*, **2**,
     215-38.
Enright, J.T. (1965). *Science, N.Y.*, **147**, 864-867.
Enright, J.T. (1975). *In: Marine Ecology* (O. Kinne, ed.), **2**, 917-944
     John Wiley, London.
Fingerman, M. (1957). *Am. Nat.*, **91**, 167-178.
Fingerman, M. (1960). *Cold Spring Harb. Symp. quant. Biol.*, **25**, 481-
     487.
Gibson, R.N. (1965). *Nature, Lond.*, **207**, 544-545.
Gibson, R.N. (1967). *J. mar. biol. Ass. U.K.*, **47**, 97-111.
Gibson, R.N. (1969a). *Oceanogr. Mar. Biol. Ann. Rev.*, **7**, 367-410.
Gibson, R.N. (1969b). *Vie Milieu*, Sér. A, **20**, 235-244.
Gibson, R.N. (1970). *Anim. Behav.*, **18**, 539-43.
Gibson, R.N. (1971). *Anim. Behav.*, **19**, 336-343.
Gibson, R.N. (1973a). *J. exp. mar. Biol. Ecol.*, **12**, 79-102.
Gibson, R.N. (1973b). *Mar. Biol.*, **22**, 379-386.
Gibson, R.N. (1975). *Proc. 9th Europ. mar. biol. Symp.*, (H. Barnes,
     ed.), 13-28, Aberdeen Univ. Press.
Gibson, R.N. (1976). *In: Biological Rhythms in the Marine Environment*,
     Belle Baruch Library in Marine Science, No. 4, (P.J. De Coursey,
     ed.), 199-213. Univ. S. Carolina Press.
Gibson, R.N., Blaxter, J.H.S. and de Groot, S.J. (1978). Developmental
     changes in the activity rhythms of the plaice (*Pleuronectes
     platessa*). This volume, pp. 169-186
Gollub, A.R. (1974). M.S. Thesis, University of Arizona, Tucson.
     57 pp.
Gollub, H. (1959). *Dt. FischZtg. Radebul*, (1959), 86-89.
Gompel, M. (1937). *C.r. hebd. Seanc. Acad. Sci., Paris*, **205**, 816-818.
Green, J.M. (1971). *Can. J. Zool.* **49**, 255-264.
Hauenschild, C. (1960). *Cold Spring Harb. Symp. quant. Biol.*, **25**,
     419-497.
Healey, M.C. (1971). *J. Zool.*, **163**, 177-229.
Hefford, A.E. (1931). *New Zealand Marine Dept., Wellington*, 32 pp.
Hickling, C.F. (1927). *Fishery Invest., Lond.*, Ser II, **9**, 100 pp.
Ishibashi, T. (1973). *Fukuoka Univ. Sci. Rep.*, **2**, 69-74.
Jens, G. (1953). *Arch. Fischwiss.*, **4**, 94-110.
Jens, G. (1954). *Arch. Fischwiss.*, **5**, 113-119.
Korringa, P. (1947). *Ecol. Monogr*, **17**, 347-381.
Korringa, P. (1957). *Mem. geol. Soc. Am.*, **67**, 917-934.
Kuipers, B. (1973). *Neth. J. Sea Res.*, **6**, 376-388.
Lang, H.-J. (1965). *Zool. Anz.*, **28** (Suppl.), 379-386.
Lang, H.-J. (1967). *Z. vergl. Physiol.*, **56**, 296-340.
Lang, H.-J. (1969). *Zool. Anz.*, **32**, 291-298.
Lang, H.-J. (1970). *Umschau*, **14**, 445-446.
Loosanoff, V.L. (1938). *Int. Revue ges. Hydrobiol. Hydrogr.*, **36**,
     170-183.
McDowall, R.M. (1969). *Tuatara*, **17**, 133-144.

Meyer, P.F. and Kuhl, H. (1953). *Arch. Fischwiss.*, **4**, 87-94.
Muus, B.J. (1967). *Meddr Danm. Fisk-og Havunders.*, **5**, 1-316.
Naylor, E. (1976). *In: Adaptation to Environment* (R.C. Newell, ed.), 393-429. Butterworths, London.
Naylor, E., Atkinson, R.J.A. and Williams, B.G. (1971). *J. interdiscipl. Cycle Res.*, **2**, 173-80.
Nishikawa, M. and Ishibashi, T. (1975). *Zool. Mag., Tokyo*, **84**, 184-189.
O'Connor, J.M. (1972). *J. exp. mar. Biol. Ecol.*, **9**, 173-177.
Palmer, J.D. (1974). *Biological clocks in marine organisms.* Wiley, London, 173pp.
Savage, R.E. and Hodgson, W.C. (1934). *J. Cons. perm. int. Explor. Mer*, **9**, 223-239.
Schwarz, F.J. and Robinson, P.T. (1963). *Progve Fish Cut.*, **25**, 151-154.
Schwassmann, H.O. (1967). *In: Atas do Simposio sobre a Biota Amazonica.* (H. Lent, ed.), **3**, 201-220.
Schwassmann, H.O. (1971). *In: Fish Physiology* (W.S. Hoar and D.J. Randall, eds.), **6**, 371-428. Academic Press, London.
Stahl, M.S. (1973). M.S. Thesis, University of Miami, 91 pp.
Taylor, A.C. and Naylor, E. (1977). *J. mar. biol. Ass. U.K.*, **57**, 273-277.
Tesch, F.-W. (1973). *Der Aal.* Paul Parey, Hamburg, 306 pp.
Tester, A.L. (1938). *Prog. Rep. biol. Sta. Nanaimo & Prince Rupert*, No. 38, 10-14.
Thijssen, R., Lever, A.J. and Lever, J. (1974). *Neth. J. Sea Res.*, **8**, 369-377.
Thomson, D.A. and Lehner, C.E. (1976). *J. exp. mar. Biol. Ecol.*, **22**, 1-29.
Thomson, D.A. and Muench, K.A. (1976). *Bull. Sth Calif. Acad. Sci.*, **75**, 198-203.
Thomson, W.F. (1919). *Calif. Fish Game Comm., Fish Bull.*, **3**, 1-47.
Thomson, W.F. (1936). *Ecology*, **17**, 158-168.
Tyler, A.V. (1971). *J. Fish Res. Bd Can.*, **28**, 1727-1732.
*Walker, B.W. (1949). Ph.D. thesis, University of California. Los Angeles.
Walker, B.W. (1952). *Calif. Fish Game*, **38**, 409-420.
Zander, C.D. (1967). *Meteor ForschErgebn.*, **D**, (2), 69-84.

* Not seen.

ON RHYTHMS OF FISH BEHAVIOUR

B.P. MANTEIFEL, I.I. GIRSA AND D.S. PAVLOV

USSR Academy of Sciences,
A.N. Severtsov Institute of Evolutionary
Morphology and Ecology of Animals,
USSR, Moscow.

In the comprehensive paper on daily rhythms of feeding and activity in fishes (Manteifel *et al.*, 1965) the basic theoretical concepts of the adaptive significance of these population patterns were presented in relation to environmental conditions; and the biological significance of the periodical, daily changes in the defensive-feeding inter-relations in fishes were revealed. The diurnal rhythmicity of feeding and activity is treated by the authors as a characteristic of fishes, reflecting the biological relationships. The food, fish and their predators - the three component links of the food chain - 'triotroph' (Manteifel, 1961) are the main components of these relationships. This concept is based on a new understanding of the nutritive base, a complicated biological phenomenon where the behaviour of consuming and consumed organisms plays a considerable part. The revealed conformities to natural laws of biological inter-relations discloses their contradictory character. Their adaptive meaning is shown by the restricted time of their activity. During 24 hours the system of fish feeding inter-relations changes radically in relation to changes of food accessibility. Changes in light intensity are related to changes in appearance of defensive-feeding behaviour of organisms. This is closely connected with the development of sensory and analytical systems and other biological characteristics of these species. Changes in food accessibility (the appearance of a new food object, its rapid increase in quantity, the appearance of new predators) completely change the character of diurnal activity in the entire food chain of a water body or in its main components.

Numerous studies, carried out at various hours of the day, show that the type of fish activity, expressed in different behavioural forms, doesn't depend on the hour of the day, but is strictly connected to the specific

experimental conditions - to their biological specifity
and to the degree of light intensity. Light provides the
necessary conditions for the existence of fishes and permits orientation for many animal species. Like any other
stimulus, especially constant ones, light plays a signal
role, preceding the biotic and abiotic changes in the
environment of organisms. Under experimental conditions
the switching on of a light can set in train daily activity rhythms during the night hours. The fishes gather in
schools, begin to hunt for food and feed on plankton.

The specific form of daily activity is the behavioural
adaptation of animals to their environment. The more or
less favourable periods for discovery and capture of food
determine the daily rhythms of feeding in fishes. The gradual decrease in the light intensity as night sets in and
its increase in the early morning hours serve as signals
of change in biotic environmental conditions. This is reflected in the behaviour of fishes from different ecological groups, belonging to different links of the food
chain of the water body. Consider, for example the important biological phenomenon of migration.

Drifting migration is characteristic for early fry.
This is an important link in the life cycle in many fishes.
These migrations appear in fry movements, on the whole,
downstream, from breeding areas to foraging areas. An
adaptive significance of drifting migration is determined
by increasing food requirements (feeding) of fry and, so
in evolutionary terms, leads to increase of abundant
species.

Passive drifting migrations are the most widespread
type of fish movement at this stage, especially among prelarvae, larvae and early fry. The potential for passive
migrations, apparently, exists for a variety of fish class
members. But the realization of these possibilities is
accomplished in conformity with the ecology of the species
and with the specific conditions of their habitats. The
greater part of this belongs to species specific complexes
of innate behavioural responses. These complexes are changing during ontogenesis in each fish species, resulting in
migration or localization in a current at specific growth
stages. The behavioural complexes leading to displacement
of fish into the current are the active elements, by means
of which the drifting migrations are achieved.

Fish behaviour, including feeding and locomotor activities, is determined in many respects by the degree of the
photoperiod in a given geographical zone during the period
of investigation. In middle and low latitudes the light
intensity may change by a factor of 10^7 precluding fish
vision at night, while in the circumpolar region the light
intensity changes only by a factor of 10^2 in the upper
water layers in June and early July. Thanks to this, light
intensity allows fishes to find food objects and to orien-

tate by means of their sight. The study of fish fry (*Clupea harengus maris albi* Berg., *Gasterosteus aculeatus* L., *Ammodytes hexapterus Pallas*) in the White Sea during summer, showed that, in spite of the polar day conditions, and the absence of a daily period of darkness, the 24 hours behavioural rhythm exists only in some fry. Changes in behaviour of fish fry were caused by changes of biological situation requiring as a rule, rapid behavioural response. In the day time herring (20-40mm) occur at some distance from the shore, and what's more, at some distance from the water surface. They manoeuvre in schools, avoiding the 'upper' and 'lower' predators. At night, the herring fry approach quite near to the shore in the shallow water. They remain in dense accumulations, often between *Fucus*. Naturally, during this period they don't feed. It can be seen how these accumulations are pursued by large *C. harengus* and *Osmerus eperlanus* (L.), which show the feeding activity at this period of 24 hours. At these hours on the water surface the light intensity exceeds 100 lux, but at some depth the twilight comes and the co-ordination of school manoeuvres grows weak, creating a favourable situation for the successful hunting of predators. The fry become inaccessible along the shore, where the larger fishes can't approach. Thus horizontal migrations of the herring fry are the main characteristics of their daily behavioural rhythmicity. Apart from the short interval at night, the diminution of feeding by herring fry is observed only after the complete filling of the stomach by food.

G. aculeatus fry move down from the spawning area to the sea and spend the whole of June and early July in shore among the weed. These fry show the day-time activity (both feeding and schooling) throughout 24 hours. At this period of the year, the light intensity on the water surface does not fall below 500 lux. Feeding in *G. aculeatus* fry is discontinued only after the maximum filling of the stomach.

While the herring and stickleback fry inhabit the upper, well illuminated water layers, the sandeel also a very common fish of the White Sea, inhabits the deeper layers close to the bottom and prefers sandy soils. In the White Sea sandeels are active throughout 24 hours during the summer months. At the same time in the Baltic Sea the related species move into the sand at evening when twilight comes and stay there until daybreak, thus changing their environment rhythmically. In the Arctic summer, the sandeel school goes into sand only under extreme circumstances (for example, when unable to avoid a predator, poor oxygen conditions, or being pulled out of water). But during continual 24 hour schooling activity the sandeels rest in turn, going into the sand. In the summer, under artificial darkening in tanks, sandeels go into the sand at 0,1-0,01 lux. This experiment emphasises once more the direct behavioural response of fishes to changing light intensity. In spite of the continuous 24 hour locomotor activity, in July the sandeel has a very precise feeding period, which occurs during the day. This takes place because the sandeel being the mass species in the coastal waters, attracts the attention of the numerous schooling predators (the large *O. esperlanus*, herrings, cod), which in the summer time feed near the coasts. Defensive schooling behaviour excludes simultaneous feeding, which occurs at the dispersion of the school and at the predominance of other activity. During the evening and at night the light intensity at the depths where sandeels live, is sufficient for them to see food organisms, but the decreased distance of visual perception and the decreased acuity make the fish more fearful. The fishes are grouped in dense schools, and feeding ceases. Analysis of the predator stomach contents shows that during the night they are feeding on sandeels. The defensive behaviour of the sandeel, as any adaptation, is relative.

Thus, during the same summer months the fry of the more widespread fish species in the White Sea have different types of feeding activity during 24 hours (although their principal feeding objects coincide). The point is that the spawning time in those three fish species differs, and, therefore, their degree of preparedness for the exhausting first months of the year, is different. Stickleback fry hatch later than the other species. These fry feed throughout 24 hours. This persistence changes at the end of July. In the darkness they switch from feeding on plankton to feeding on larval and adult Chironomidae. Thus, their feeding rhythmicity is expressed by the changing quality of the food objects whose accessibility changes during the 24 hours. Sandeel fry, which hatch in the winter, already have a considerable fat content in the summer. In this connection, their feeding intensity is lower than in sticklebacks and herring and during a 24 hour period these fry have a considerable interval in their feeding. In this respect, herring fry are intermediate.

We should emphasise that the character of daily feeding rhythmicity isn't only a result of ecological relationships of organisms at a given stage in their life, but is also a reflection of previous events and preparation for the coming season.

As noted previously, the daily feeding rhythmicity can be changed artificially under experimental conditions. With natural changes of light intensity in an aquarium, fish fry, as in the wild, change their behaviour. From the morning to evening, when the light intensity exceeds 0.1 lux, *Rutilus rutilus* (L.) and *Leucaspius delineatus* (Heck) fry stay at the water surface, feed and swim in schools. When the light intensity is lower than 0.1 lux, the fry are found in the middle and lower water layers,

dispersed, flattened against the aquarium walls and stationary. The fry spend the dark period in tactile contact
with the aquarium walls or with vegetation, coming to the
surface in the morning. This behavioural pattern was disturbed by the introduction of a nourished predator (*Esox
lucius* L.) into the aquarium. The specific defensive behaviour of fish becomes evident during the first minutes
of the experiment. The fry school aggregates, moves away
from the surface into the middle waterlayers, manoeuvring
between the surface and the bottom. This school ceased to
feed, and did not come to artificial light at night. Thus,
the continual defensive stereotype of the fry behaviour
disturbed the daily behavioural rhythmicity. In other experiments the defensive breaking of feeding behaviour in
small perch in the dark was stopped by keeping them without predators for two months. These individuals began to
hunt and feed in complete darkness although in the wild
their feeding activity ceased at twilight.

We treat daily rhythmicity of the vertical distributions
in cyprinid fish fry (that is their nocturnal migration
into the lower water layers into the weed or on to the
bottom, after a dispersion of the school), on the one hand
as an adaptation keeping the fry in a habitat with good
living conditions, and, on the other hand, as defensive
behaviour during the hours of darkness. As soon as the
fish can achieve visual orientation during the twilight,
their drift downstream is accomplished. The prelarvae and
early larvae can't resist the current and they drift all
day and night. This passive drift is typical of sturgeon
prelarvae and larvae, of salmon larvae as they emerge from
the spawning redd, of the first larval stages in many per-
cids, cyprinids, carp, clupeids and in other fish species.
Concerning the pelagophil fishes, their passive drift begins at the spawn stage.

During growth and development and the increase of their
swimming activity, the significance of such drift in the
passive migrations is decreased. In a lentic environment
with sufficient water transparency, the fry tend to keep
near the shore during daylight. As a rule, the current
speed here corresponds to their swimming ability. Here,
the larger the fish, the greater the velocities they can
withstand. As a rule, the fry are orientated against the
flow and move upstream or maintain station at the certain
defined areas. Such orientation is kept until twilight
comes. This orientation breaks down only when the light
intensity decreases below the threshold for the visual
mechanism of rheoreaction (the optomotor reaction). Then,
the fry drift with the current. When daybreak comes, the
fish can orientate again and their positions in the streams
are restored.

In reservoirs with extremely low transparency (4-15 cm
by the Secchi disk), this rhythmicity isn't really related

to light intensity. It must be noted that an artificial
increase of turbidity leads to the downstream migration
of fry during daylight.

In reservoirs where the turbidity is sufficient for
visual fish orientation, the daily rhythms of drifting
migrations are well marked not only in cyprinid fish, bu
also in Clupeidae, Percidae, Siluridae, Cobitidae, Gobii
dae. (Pavlov, 1966, 1970). This drift rhythm is typical
not only for the migratory or semi-migratory, but also
for the native resident fishes.

The timing of twilight depends on geographical locatic
therefore, the time of the start of the fry migration ca
change accordingly. The nearer to the North, the shorter
is the summer night and also the diurnal period of the p
sive drifting migrations.

In the experiments on cyprinid fish fry we are able t
establish that, the larger the fry, the later (during th
24 hours) and at the lower light intensity their migratic
begin. It is connected to the fact that during growth, t
threshold values of light intensity for the optomotor reaction (the visual mechanism of rheoreaction) become lower. So, for example, the vobla (*Rutilus rutilus caspicus*
fry at the length: 6-9, 10-15, 15-26 mm has the following
values of light intensity threshold: 0.1, 0.01, 0.001 lu
On the other hand, the larger the fry, the sooner it sto
migrating. In the night, the light intensity near the sh
and in the surface layers is sufficient for visual orientation. So, for example, on moonlit nights the light int
sity at the water surface reaches about 0.1 lux, but dur
the clear star nights about 0.001 lux. The level of light
intensity is sufficient for orientation in fish fry of a
defined size. When the dark adaptation process is complet
ed, the fry migration intensity decreases.

Thus, the larger the fry, the narrower is the time
interval during which the drifting migration is possible.
At later growth stages the insignificant drift by the
current at night is compensated by the day-time fry movements against the flow.

When the tactile mechanism of orientation in the current begins to function, the fry appear to be able to
orientate against the flow in darkness. In many Teleostei
this moment is related to their ecology and, apparently,
defines the end of the drifting migration period. Obviously, in many bottom fishes the tactile orientation mech
anism appears at the earlier growth stages. In all cases,
the drifting migrations in such fishes as *Gobio*, *Nemachil
us*, *Gobius*, are finished, as a rule, earlier than in othe
fishes. The migrations of a sturgeon are related to fry
ascent from the bottom to the mid-water.

During our experiments, the individual vobla specimens
begin to orientate against the flow in darkness at length
of 23-30 mm (roach fry - about 30 mm, perch fry - 38-39

m). The orientation in the majority of fishes in our
experiments come later: vobla - at lengths of 37-38 mm,
roach - 44-45 mm, *Leucaspius delineatus* - 31-33 mm, perch
45-50 mm. Undoubtedly, the better the tactile orientation
conditions are, the smaller the size at which the fish's
drift by the current can cease. When the tactile mechanism
of rheoreaction begins to function, and the fish can orien-
ate against the current, the passive drifting migrations,
as a rule, cease or abruptly decrease, and they aren't of
mass character.

Because the passive drifting migrations occur in the
dark, the weak agility of fry during this period and the
dispersion of fry along a river section play an important
role in the defence of drifters against their destruction
by the twilight-night predators.

The daily rhythmicity of the adult fishes migrations
during the prespawning period can be divided into four
groups: day-time, night-time, twilight and all-day-and-
night (with maximum during the night hours).

Herring, bream, vobla, *Pelecus*, *Scardinius* and perch
belong to the *day-type of migrations*. The change of intensity of migrations in these fishes clearly corresponds to
the change of light intensity, increasing during their
ascent in the morning. The main migration occurs at the
light intensity of about 10,000 lux. In the twilight the
migration is diminished, and at night at a light intensity
of about 0.01-0.001 lux - almost completely ceases.

The night-type of migration (0.01-0.001 lux) is shown
by the pike-perch (*Stizostedion* spp.) and the freshwater
catfish (*Siluridae*). The maximum intensity of these fish
migrations occurs at the light intensity of about 0.001
lux. The presence of the moon can reduce the intensity of
their migrations. *Lamptera* can be included in this group
of migrants. The movement of *Lampetra* is said by some
authors to be observed only at night.

We found that *the twilight type of migration* is shown
in the salmon, *Cyprinus* and *Blicca*. The migration in these
fishes occurs in wide ranges of light intensity - from
0.001->1,000 lux. Apparently, it would be correct to divide the fishes of these types into twilight-day (*Cyprinus*,
salmon) and twilight-night (*Blicca*) types. During cloudy
weather, when the light intensity is significantly decreased the twilight-day type fishes migrate also by day. That
is light intensity, and not a particular time of day, is
the principal condition of migration. So, for example,
salmon migration takes place, on the whole, at the light
intensity about 5-15,000 lux. If this light intensity
takes place in the day, the migration occurs also during
these day-hours.

The sturgeon, *Acipenser*, *Vimba* and apparently, the
sterlet (data on the sterlet are scanty) belong to *the
all-day-and-night type of migration*. The maximum of these

migrations occurs, on the whole, during the night, but
the intensity of migrations is also high during the day.
The decrease of the intensity of migrations during daylight in sturgeon and *Acipenser* in the Volga delta, usual
ly, doesn't exceed 50%. Near Volgograd, where the water
transparency is 2-3 times higher than in the Volga delta,
the night migration peak was considerably more clearly
expressed. These data suggest a close relation between
the all-day-all-night type of migration and the night
migration type. During the season, the daily migration
dynamics in sturgeon (as, also, in certain other fishes)
changes, corresponding to the seasonal alternations of
the light intensity rhythm (the night period - the period
of migration is increased towards the autumn).

Considering the relationships between the type of diur
al rhythmicity of fish upstream migration and the characteristics of orientation in the current, it can be mentioned that fishes in which visual mechanisms dominate
in orientation can be attributed to *day-type* and to *twilight type*. The bottom fishes, on the whole, with tactile
mechanism of orientation (moreover, the lateral-line organs are well developed in these fishes - freshwater catfish, pike-perch, *Vimba*) can be attributed to the *night
type* and to the *all-day-and-night type*. Thus, the characteristics of orientation are of great importance in the
formation of diurnal rhythms of spawning migration in
fishes. However, it would be an oversimplification (and,
in case of tactile fishes, incorrect) to explain the
daily migration rhythmicity only by the characteristics
of orientation in a current. The adult individuals of
many species (particularly of bottom species) have the
capacity for orientation in a current at any hour. Therefore, it must be considered that the daily rhythmicity of
the upstream migration is formed by both the behavioural
type and the locomotor activity of fishes. It's not acci-
dental, that the defined types of spawning migration rhyt
micity are analogous to those types, which were determine
from the study of the diurnal dynamics of the food-defen-
sive inter-relations in fishes from Rybinskoe reservoir
(Manteifel *et al.*, 1965) and which were described by
Manteifel (1970, 1973).

The strength of influence of one or other environmenta
factor alters the speed of the behaviour response, and
this depends on the relative biological significance of
these factors during the life of fishes. The artificial
light field not only breaks the rhythmicity of daily
activity in fishes, but also alters the direction of reaction and disorientates the fishes in both time and
space. This isn't surprising. Light is a rich carrier
of biological information, the necessary condition of
life, particularly in early fry of fishes. In the pressur
gradient apparatus the fry of *Rutilus rutilus* (L.) (lengt

17-20 mm), *Alburnus alburnus* (6-10 mm), and *Perca fluviat-ilis* L. (10-24 mm) move to the light in the direction of the zone of decreased pressure. Besides, the fry move at the same speed as in their movement to higher water layers (the zone of the lower pressure). Thus, only the light is a guiding factor on the surface during daily vertical migrations of fish fry. The possibility of both the directional changing of activity and of breaking of the daily rhythm are widely used in management of fish.

The changes of the biological situations in interspecific relations between fishes occur instantaneously and demand rapid behavioural responses. The seasonal influences of many abiotic environmental factors occur over prolonged periods and affect metabolic processes, resulting in profound physiological changes in the organism. The seasonal changes of the physiological state in fishes are based on the complex influence of rhythmically repeated light and dark periods (of different time) and also of defined temperature. Fish life depends on the manifestation of characteristic activity at the appropriate season of the year. Fish response to light in photogradient conditions can serve as its test.

It is possible under experimental conditions to separate the influence of a rhythmic photoperiod (with differing proportions of light and dark periods within 24 hours) and the influence of temperature on fishes. The studies were carried out on *Leucaspius delineatus* (Heck.). Two groups of fish, held at differing photoperiods (17L:7D, 7L:17D) each 24 hours during one month at the same temperature (17-18°C) showed different responses to light. The *L. delineatus* experiencing short-days avoided the illuminated zone and were actively concentrated in the darkest part of the photo-gradient. The long-day fishes showed positive responses to the illuminated zone. Two other groups of fishes were held under continuous light (700 lux), but at different temperatures and behaved as the fishes, held under at the short-day conditions. The behavioural response of the fishes to the short-day or to the low temperature conditions, was considered by us an adaptation to the winter season: moving away from the zone of abnormally high exchange' into the 'zone of energy preservation'. In this case, light can be treated as a signal of the definitive complex of ecological conditions, favourable for the summer season and unfavourable for fishes in the winter period. This seasonal change in photo-reaction in fish will influence the success of trapping methods which use light as an attractant.

References

Manteifel, B.P. (1961). *Trudy In-ta morfologii zhivotnykh AN SSSR*, **39**, 5-46.
Manteifel, B.P. (1970). *In: 'Biologicheskie osnovy upravleniya*

povendeniem ryb'. Izd. Nauka, Moscow, 12-35.
Manteifel, B.P. (1973). *In: Trofologiya vodnykh zhivotnykh,* Izd.
   Nauka, Moscow, 85-94.
Manteifel, B.P., Girsa, I.I., Leshcheva, T.S. and Pavlov, D.S. (196:
   *In: 'Pitanie khishchnykh ryb i ikh vzaimootnosheniya s kormovymi
   organizmami,* Izd. Nauka, Moscow, 31-81.
Pavlov, D.S. (1966). *Voprosy ikhtiologii,* **6**, 528-539.
Pavlov, D.S. (1970). *Optomotornaya reaktsiya i osobennosti orientat:
   ryb v potoke vody,* Izd. Nauka, Moscow.

LOCOMOTOR ACTIVITY IN WHITEFISH - SHOALS
(*COREGONUS LAVARETUS*)

KARL MÜLLER

Department of Ecological Zoology,
University of Umeå,
S-901 87 Umea,
Sweden.

Introduction

Few ethological phenomena in the field of fisheries biology have received so much attention as the formation and disbanding of shoals. The work performed has resulted in considerable experimental advances, but at the same time has produced a great deal of speculation. All factors, except one, that could conceivably elicit the formation of shoals have been incorporated in the various hypotheses put forward, the exception being the daily light-dark changes. Thus, light intensities have been studied by Hunter (1968), temperature by Breder (1959) and chemical composition of the water by Hemmings (1966) and other factors by Keenleyside (1955) and Shaw (1960), just to mention the most important studies. Siegmund (1969), however, in connection with field studies, reported that young perch (*Perca fluviatilis*) tend to form shoals at sunrise and disperse at sunset. After developing the necessary experimental techniques we have been able to continuously record the movements of dayactive whitefish (*Coregonus lavaretus*) throughout an entire year-cycle. Both shoals of 100 individuals and isolated fishes have been observed. In this way it has been possible to analyse the factors that regulate shoal-formation in this species.

Material and Methods

The investigation was carried out at the Messaure Ecological Research Station (66°42'N, 20°25'E), where the fishes were kept in an experimental tank. The tank which had a total volume of approx. 4 m³ was divided into eight sections, each with a length of 1 m, and a height of 0.50 m. There was a continuous flow of water through the aquarium, with the water being completely replaced 6-8 times per 24 h period. The bottom of the aquarium was covered with sand, gravel and small stones. The water flowed

into the aquarium at four points at one end; the water
current so produced had a velocity of 0.10-0.15 m/sec.
Four photocells were placed at equal distances around the
aquarium. Light-beam interruptions were recorded on an
Elmeg time printer and were summed over hourly intervals.
Shoal formation and disbanding were observed directly too
All experiments were performed on fishes of the ageclass
(Group 1+) on fishes which were 12-15 cm long.

Results

The Diel and Seasonal Locomotor Activity Patterns of
Solitary Fishes and Shoals of Coregonus lavaretus.

   Figs. 1 and 2 show the locomotor activity of a single
fish and of a shoal respectively, represented as monthly
means, for a period of one year. Except for the months of
June and July, the figures are identical. Disregarding
the months around midsummer when light intensity never
fell below the 5-lx threshold, the solitary fish and
the shoal always followed this value in connection with
the onset and end of activity.
   Fig. 3 shows the activity of a shoal of 100 fishes
for five consecutive days in winter, spring, summer and
autumn during 1970, in the experimental tank at Messaure.
In this series of figures it is easy to notice the dependence of the shoal's locomotor behaviour on the changes
in the natural light-dark period. The onset of swimming
activity corresponds exactly with the formation of the
shoal, and the decline in activity is similarly simultaneous, with the break-up of the shoal. Between September
and May it is not possible to detect any swimming activity during the night. With increasing water temperatures,
around the end of May, the general level of activity is
enhanced, but the diel periodicity is maintained. This
is, however, lost for the short period (19-28 June at
Messaure - at the Arctic Circle) when the sun never sinks
below the horizon. At this time the shoal displays an
arhythmic behaviour. The arhythmicity is even more pronounced in solitary fishes. During the period when the
light intensity never fell below 5-lx, no diel periodicity was observed among solitary fishes (8 of which
were tested in parallel). It thus appears that the shoal
is less influenced than are solitary fishes by extreme
light conditions such as those prevailing north of the
Arctic Circle.

Cumulative Activity of Solitary Fishes and Shoals of
Coregonus lavaretus

   Fig. 4 shows the cumulative activity as a mean for
8 solitary fishes per 12 months period. In Fig. 5 the
corresponding values are shown for the shoal. The continuous line represents the monthly means for the water temperature. The cumulative sum for the solitary fishes varies

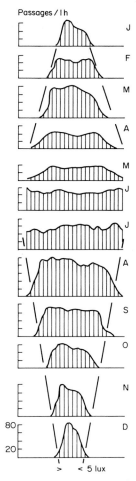

Fig. 1. Daily distribution of locomotor activity (monthly mean) of a single whitefish (*Coregonus lavaretus*) during one annual period at the Arctic Circle (Messaure).

between 285 photocell passages/day in January, and 1950 passages/day in July. The corresponding values for the shoal were 1450 and 29,500 passages/day respectively. The cumulative activity sum is to a very high degree temperature-dependent. Despite the long daylength during April-May at the Arctic Circle, the activity increases steeply only when water temperature starts to increase around the end of May. A similar correlation between decreasing activity and falling temperatures is seen in the autumn.

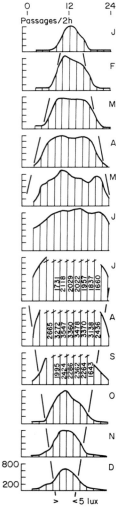

Fig. 2. Daily distribution of locomotor activity (monthly mean) of a shoal of 100 whitefishes during one annual period at the Arctic Circle.

The Duration of Activity of Solitary Fishes and Shoals of Coregonus lavaretus

The daily period of activity of solitary fishes is shown in Fig. 6 and that for shoals in Fig. 7. Except for the special conditions that prevail around midsummer, the activity period and daylight period are generally more or less identical. Differences did occur however around midwinter, when the 8 solitary fishes had an average

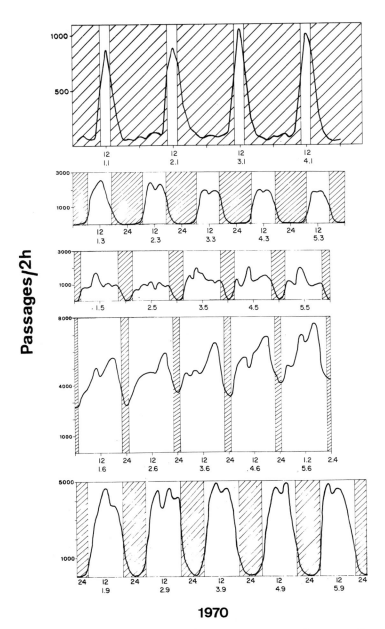

1970

Fig. 3. The course of the daily activity-rest changes of a whitefish shoal (100 fishes), shown for five consecutive days in January, March, May, June and September 1970 at the Arctic Circle.

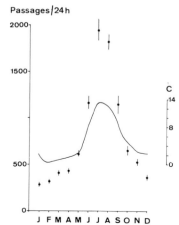

Fig. 4. The amount of locomotor activity of a solitary whitefish (mean of 8 solitary fishes) in the course of the year (monthly means) at Messaure.

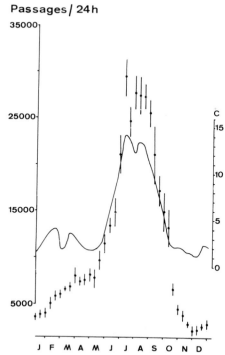

Fig. 5. The amount of locomotor activity of a whitefish shoal (100 fishes) in the course of the year at Messaure.

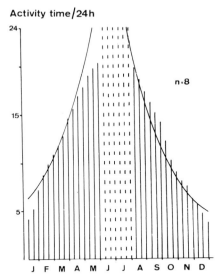

Fig. 6. Duration of activity in the 24 h period (10-day means) of a solitary whitefish (mean of eight solitary fishes) in the course of the year at Messaure. Straight solid lines = activity time. Dashed lines = summer arhythmicity. Curves = 5 lx thresholds.

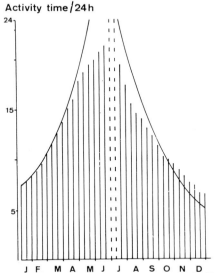

Fig. 7. Duration of activity in the 24 h period (10-day means) of a whitefish shoal (100 fishes) in the course of the year. Key as in Fig. 6.

activity period of 4.25 h whereas the shoals never had an activity period shorter than 6.50 h.

Formation and Break-up of the Shoal

The investigated shoal of 100 fishes spent the nights irregularly dispersed throughout the aquarium. Orientated towards the water inlets they compensated for the current by slow movements of their pectoral fins. At first light small shoals of 5 to 10 individuals are formed that swim against the weak water current. Gradually more and more fishes join the shoal which finally becomes 2-2.5 m long and very compact. The shoal remains intact all day and in the evening the reverse procedure takes place with fishes leaving the shoal one by one and taking up the night-time resting position. Formation and dispersal take place for about 20-30 min. in August near the Arctic Circle (Fig. 8).

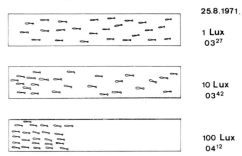

25.8.1971.

1 Lux
03^{27}

10 Lux
03^{42}

100 Lux
04^{12}

Fig. 8. Schematic presentation of the formation of a whitefish shoal in relation to the increase of the light intensity in the morning.

A similar night-time behaviour has previously been described by Fabricius and Lindroth (1953) from laboratory experiments performed at the Hölle Laboratory in northern Sweden: 'The school of young whitefishes dispersed and its former members moved about singly in all parts of the tank during the night'.

Discussion

Improved equipment has made it possible for us to follow the behaviour of shoals and individual whitefishes continuously over long periods of time. The results presented here show clearly that the formation and dispersal of shoals are intimately connected with the daily alternations between daylight and darkness. In the experimental arrangement I have described, the fishes were exposed to a weak water current which is not always the case in their natural lacustrine surroundings. At the same time we performed experiments in a vertical aquarium which did not have any water current (Eriksson and Kubicek, 1972), which gave essentially the same results. Thus we can conclude that regardless of habitat the formation and break-up of shoals of whitefishes is primarily determined by the alternation in the diel light-dark conditions. This does not, of course, exclude the possibility of some social factor functioning secondarily as a zeitgeber and tending to hold the shoal together. An indication that such a factor is, in fact, involved was obtained from experiments with artificial light-dark change-overs performed with both shoals and solitary fishes. Under artificial illumination conditions with a 12:12 h light-dark cycle the activity period of both the solitary fishes and of the shoals closely synchronized with the periods of light. When, however, the solitary fishes were transferred to complete darkness (DD) their activity immediately became arhythmic. The shoals, on the other hand, remained rhythmic and maintained their phase position for up to 10 days after being transferred to complete darkness. It seems highly probable that despite the total darkness the fishes were in some way able to remain in contact with each other. By what means this communication takes place - whether it is chemical or acoustic - remains to be seen, and is a suitable problem for coming research projects.

Acknowledgement

The work is supported by the Swedish Natural Research Council, the Max-Planck-Gesellschaft zur Förderung der Wissenschaften and the Deutsche Forschungsgemeinschaft.

References

Breder, C. (1959). *Bull. Am. Mus. nat. Hist.* **177**, 395-481.
Eriksson, L.O. and Kubicek, F. (1972). *Fauna och Flora* **67**, 24-28.
Fabricius, E. and Lindroth, A. (1953). *Rep. Inst. Freshwater Res. Drottingholm* **35**, 105-112.
Hemmings, C.C. (1966). *J. exp. Biol.* **45**, 449-464.
Hunter, J.R. (1968). *J. Fish. Res. Bd. Canada* **25**, 393-407.
Keenleyside, M.H. (1955). *Behaviour* **8**, 183-247.
Shaw, E. (1960). *Physiol. Zool.* **33**, 79-86.
Siegmund, R. (1969). *Biol. Zbl.* **88**, 295-312.

ACTIVITY RHYTHMS IN GYMNOTOID ELECTRIC FISHES

HORST O. SCHWASSMANN

Department of Zoology, University of Florida, Gainesville, Florida 32611

and

Instituto Nacional de Pesquisas da Amazonia, Manaus, Amazonas, Brazil

Circadian rhythms in fish, manifest as periodic changes of activity and rest, have been known for a long time, although the endogenous nature of these rhythmic phenomena in fish was demonstrated only recently (Lissmann and Schwassmann, 1965; Schwassmann, 1971a).

The idea of utilizing the continuous electrical signals produced by gymnotoid electric fish for monitoring their activity rhythms came during field work in 1960 in the Amazon estuary (Egler and Schwassmann, 1962) when we encountered many of these interesting fishes and could listen to their electric pulse with audioamplifiers and speakers. Approximately 40 species of gymnotoids have been described and five families can be distinguished on the basis of morphological features and electric organ discharge characteristics. All of these fish emit continuous trains of low voltage discharges of species-specific wave form and repetition rate. One group, the so-called wave species, or 'hummers' (Sternopygidae and Apteronotidae), produces rather broad signals of modified sinusoidal shape. Discharge rates of these wave species are remarkably stable and can be very high in some (1,800 Hz). Fish of the other group, so-called pulse-species or 'Buzzers' (Gymnotidae, Electrophoridae, and Rhamphichthyidae), emit brief spikes of specific form, mono- to polyphasic, that are separated by long intervals. In most, but not all of this group, the pulse repetition rate shows sudden increases when the fish are disturbed, or while they are swimming actively, and during feeding. Many exhibit a relatively low resting frequency during the day alternating with a higher discharge rate at night. A few have accessory electric organs in the head region which they discharge in synchrony with the principal electric organ, usually located in the tail and posterior body. One species, the famous electric eel, is capable of producing a powerful

shock of up to 500 volts by means of a very large main
electric organ.

The continuously emitted electric pulses, in conjunction
with a great number of very sensitive electroreceptor organs that are part of the lateral line system, and are distributed over the body surface, form the basis of an active
electro-orientation and electro-communication system. This
sense detects impedance changes in the near field caused
by the presence and the movement of objects, or other fish,
that are of different electrical conductance to the fresh
water. All gymnotoids are active at night and they spend
the day sheltered in plants or crevices, sand, or muddy
substrate. The adaptive significance of their electro-
sensory ability, enabling an active night life, was recognized by Lissmann (1961). One species, *Gymnorhamphichthys*,
remains buried in the sand all day, where sufficient oxy-
genation is provided by the water flowing through the
spaces in the coarse sand of the streams which these fish
inhabit. The sand-burrowing habit was known already to
Field Marshal Rondôn (Ribeiro, 1920). While engaged in
field work with populations of this species in 1964, we
found that two frequency levels were associated with the
states of activity and rest (Lissmann and Schwassmann,
1965). During daytime when these 'sandfish' are resting
in the sand they discharge their electric organs at a low
rate of 10 to 15 Hz, showing occasional slight frequency
increases. When they emerge after sunset and swim freely,
discharge rate of electric pulses is from 60 to slightly
over 100 Hz with rare and very brief increases to close
to 200 Hz. Pulse frequency is maintained at the high level
until the fish re-enter the bottom substrate during late
night or early morning hours.

Gymnorhamphichthys specimens exhibit the same behaviour
when they are kept in the laboratory inside individual
aquaria provided these contain sand on the bottom. Moreover, the rhythmic pattern of swimming actively and resting in the sand continues at its own endogenous free-
running period when light/dark changes are eliminated.
As can be seen in Fig. 1, the activity rhythm in constant
dark (less than 0.5 lx) has a period of about 23.5 h in
fish *A* and *B*, while the pattern appears less precise and
the period not much different from 24 h in specimen *C*.
In these experiments the free-running period and the ratio
of activity time to rest time was found to depend on the
intensity of constant light. Figs. 2 and 3 show examples
of records in such 'constant' conditions of *Gymnorhamphichthys* and *Hypopomus* spp. As predicted by Aschoff's rule
(Aschoff, 1958, 1960), the period length in these night-
active animals increases with increasing light levels,
while at the same time the duration of activity becomes
relatively less.

Continuing field studies investigated reproductive

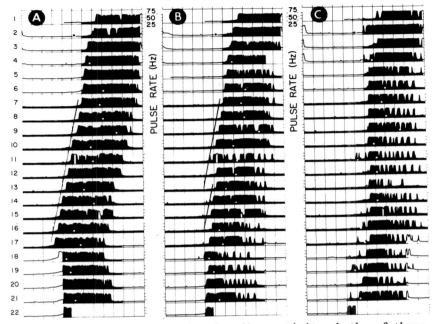

Fig. 1. Selected records of the circadian activity rhythm of three specimens of *Gymnorhamphichthys*. 24 h segments of recorded electric pulse rates are mounted in vertical sequence. The frequency record is solid during dark and open during light. The circadian period in free-running constant dark conditions (0.5 lx) from day 7 to 17 can be measured by the slope of the auxiliary lines. Open circles indicate feeding times. (Modified from Lissmann and Schwassmann, 1965).

timing patterns of populations of gymnotoid fish with the tentative goal in mind to ascertain if the demonstrated precise circadian rhythms might somehow be causally involved in a long-term advance timing process of gonadal development. Some preliminary findings have been published elsewhere (Schwassmann, 1976); and it became apparent that at least the sandfish, *Gymnorhamphichthys* is compensating for the absence of reliable annually periodic timing cues, as for example changing photoperiod, by extending its reproductive activities over a great portion of the year, and becoming a partial spawner with four to five successive spawning bouts. Synchronization of spawning activity was observed within populations of these fish but no detailed behavioural or other mechanism could be demonstrated as yet to explain how this mutual synchronization is accomplished (Schwassmann, 1976; and this volume).

Several species of *Hypopomus* and *Hypopygus*, many of them not yet described in the taxonomic literature, were collected and their electric organ discharges were monit-

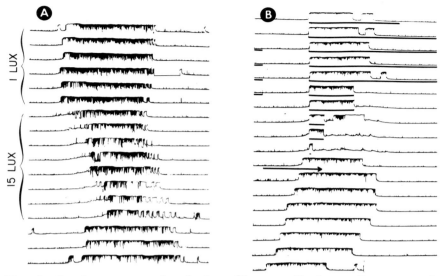

Fig. 2. Continuous records of electric organ discharge frequency of
two *Gymnorhamphichthys* showing the dependence of the circadian period
and of the activity/rest ratio on the intensity of continuous light
in *A*, and entrainment by dark pulses of different duration, also a
brief free-run in constant dark, in *B*. Darkness is indicated as a
horizontal bar under the records in *B*. (Modified from Schwassmann,
1971b).

ored in order to ascertain if these species also exhibited
frequency differences that were correlated with alternating states of activity and rest. An additional aim was to
utilize the characteristic electric pulse rates and their
variabilities, as well as any day/night changes, as a
taxonomic tool in distinguishing between these many closely
related species of *Hypopomus*. The preliminary data are
summarized in Table 1 where *Gymnorhamphichthys* is included
for reasons of comparison. Specimens of *Hypopomus* species
x were used for producing the records shown in Fig. 3.
Inspection of the Table makes it apparent that most species
of *Hypopomus* show at least some day/night differences in
discharge rates, as well as 'bursting', like *Gymnorhamphichthys* and also *Gymnotus* spp. (not shown). Bursting consists of very sudden frequency rises and slower, more
gradual declines in rates, mostly in response to external
disturbances, rarely also apparently spontaneously. *H.
occidentalis* responds to disturbances with a sudden level
increase, the subsequent decline in pulse rate takes several minutes rather than seconds as in bursting. *H.* sp.
b is different in that it exhibits only slow but almost
constant increases and decreases of pulse rates which
have about the same rather long time constants. They do

TABLE 1

List of several collected species of Hypopomus, new species identified by letter, and their respective electric organ discharge (EOD) rates and other characteristics. Hypopygus and Gymnorhamphichthys are included for comparison. Single values mean that only one specimen was used for recording. When ranges are given, several fish were used.

Species	Locality	EOD frequency low (day)	EOD frequency high (night)	other EOD characteristics
Hypopomus artedi	Manaus	8 – 11	28 – 43	bursting, day/night levels
H. x	Manaus	2 – 3	20 – 30	bursting, day/night levels
H. y	Belém			bursting, day/night levels
H. z	Manaus Rondonia	14 30	40 58	bursting, day/night levels
H. r	Rondonia	66	66	no variation noticed
H. q	Rondonia	12	25	some bursting, day/night levels
H. n	Rondonia	25 – 27	32 – 63	bursting, day/night levels
H. b	Belém	20 – 50	30 – 70	no bursting, gradual increases and decreases
H. occidentalis	Panama Colombia	23 – 38	70 – 96	no bursting, level shifting, small day/night difference
Hypopygus lepturus	ubiquitous	46 (60)	62 (96)	no bursting, very little day/night difference
Gymnorhamphichthys hypostomus		10 – 15	60 – 100	bursting, distinct day/night levels

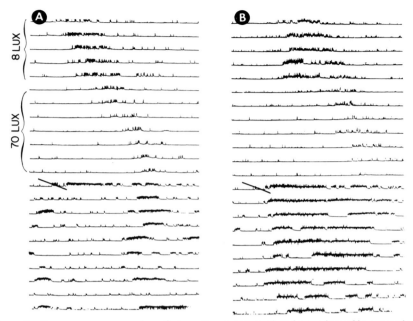

Fig. 3. Frequency records of two *Hypopomus* showing circadian periods under free-running conditions at different intensities of constant light. Activity in these *Hypopomus* recordings is seen as level increase as well as upward pen deflections and is greatly reduced at the higher light level. (Modified from Schwassmann, 1971b).

not stay with one particular frequency for very long; no fast bursting-type of increases were observed. *H.* sp. *r* was found to maintain a steady rate day and night, and resembles the wave species in this respect; also, its pulse is rather broad, but it is polyphasic. *Hypopygus lepturus* lacks the bursting response but shows some level differences between night and day, in contrast to the closely related *Steatogenys* spp. which exhibit almost no noticeable day/night differences.

Many of the gymnotoids listed in Table 1 have a wide range of distribution in the South-American tropics. It is hoped that continuing field studies will eventually result in pointing out geographic differences in the times of spawning, as well as regionally specific environmental factors which, perhaps by some involvement of the known precise circadian rhythms, are capable of entraining the annual rhythm of gonadal ripening and spawning in these and other fish living in the tropics.

References

Aschoff, J. (1958). *Z. Tierpsychol.* **15**, 1-30.

Aschoff, J. (1960). *Cold Spring Harbor Symp. Quant. Biol.* **24**, 11-28.
Egler, W.A. and Schwassmann, H.O. (1962). *Bol. Mus. Paraense E. Goeldi* NS **1**, 2-25.
Lissmann, H.W. (1961). *In: Bioelectrogenesis*, (C. Chagas and A. Paes de Carvalho, eds.), Elsevier, Amsterdam, pp. 215-226.
Lissmann, H.W. and Schwassmann, H.O. (1965). *Z. vergleich. Physiol.* **51**, 153-171.
Ribeiro, A. de Miranda (1920). *Zool. Hist. Nat. Ann.* (5) 58 1-15.
Schwassmann, H.O. (1971a). *In: Fish Physiology* **6**, (W.S. Hoar and D.J. Randall, eds.), Academic Press, New York. pp. 371-428.
Schwassmann, H.O. (1971b). *In: Biochronometry*, (M. Menaker, ed.), Nat. Acad. Sciences, Washington, D.C., pp. 186-199.
Schwassmann, H.O. (1976). *Biotropica* **8**, 25-40.

THE SIGNIFICANCE OF LOCOMOTOR AND FEEDING RHYTHMS IN THE REGULATION OF POPULATION BIOMASS AND PRODUCTION OF THE UPLAND BULLY, *PHILYPNODON BREVICEPS* STOKELL

D.J. STAPLES

*C.S.I.R.O., North Eastern Regional Laboratory,
P.O. Box 120, Cleveland, Queensland, Australia.*

Introduction

In recent years, an increasing number of studies have been carried out to provide estimates of production for fish populations. These estimates by themselves, however, add little to our understanding of the aquatic ecosystem of which these populations form a part. More emphasis should be given to defining the role various factors and processes play in regulating the flow of energy through the system. At the population level, this can be achieved most conveniently by considering the factors separately for the two main components of production - numbers and weight (energy content) (Allen, 1969). However, in a fish population, where there is often a significant interaction between numbers and growth (Backiel and LeCren, 1967), the factors do not act independently, making the concept of regulation of population biomass, rather than simply population numbers alone, of considerable ecological importance.

Diel rhythms of activity and feeding influence the production of a population mainly through their effect on the growth rate. They can act either directly, through their influence on the food consumption, or indirectly, by their effect on the level of intraspecific competition within a population. This paper deals with these effects in relation to the food intake, biomass and production of a New Zealand native fish, *Philypnodon breviceps* Stokell.

Philypnodon breviceps (Eleotridae) is a small benthic fish which inhabits many streams and lakes throughout the South Island of New Zealand and the southern region of the North Island (Hopkins and McDowall, 1970). Its general life history and ecology is similar to that of *Cottus gobio* L. (Cottidae) in the Northern Hemisphere (Smyly, 1957). It has a maximum life-span of 4-5 years, females becoming sexually mature after 1-2 years and males after

2-3 years. Mature males are polygamous and territorial in defence of a nest during the breeding season (October-December). For the remainder of the year, individuals are not strictly territorial, but to maintain their position in the marked size hierarchy which exists throughout this time, each individual actively defends a small individual space of approximately 10 to 20 cm radius. The fish are basically benthic carnivores; in the study lake, young fish fed predominantly on crustaceans, but the diet changed to larger insect larvae and young fish with increasing size and age.

Methods

The study area and general methods of study have been described by Staples (1975a, b and c). Catches taken in trap nets (modified from a design of Crowe, 1950) were used as indicators of the level of locomotor activity as suggested by Stott (1970). Four trap nets were set every morning and evening during the first week of alternate months; each trap was set at a new site selected at random over the entire lake. The average catch per hour of fish taken during the day compared with that taken during the night could thus be used to describe diel activity rhythms. Seasonal catch rates were adjusted for differences in density of fish at different times of the year by calculating an activity index (AI) as:

$$AI = \frac{n}{t} \cdot \frac{a}{\hat{N}}$$

where n is the number of fish caught, t is the duration of trapping (in hours), a is the surface area of the lake, and $\hat{N}$ is the estimate of population size. Fish were aged by a combination of length frequency and scale reading techniques.

Feeding periodicities were monitored by determining changes in the food consumption rate over 4 h intervals throughout replicated 24 h periods. A sample $(\bar{n} = 40)$ was taken with a push net (modified from a design described by Strawn, 1954) every 4 h and half the fish were killed immediately on capture, while the remainder were left for the next 4 h period in a food-free container in the lake. The food consumption rate was estimated as the difference in the dry weight of stomach contents (mg/g dry weight of fish) between fish held in the container and those feeding naturally in the lake for the same period. The experiment was repeated every three months to follow seasonal changes for one year class (age 1+ - 2+) and a comparison among age groups was also made in the summer of 1970.

The production dynamics of the population were calculated by the method of Ricker (1946), i.e.

$$P = G.\bar{B}$$

where P is the production calculated over monthly intervals, G is the instantaneous growth in weight and $\bar{B}$ is
the mean biomass for the period. Population numbers were
determined from bi-monthly capture-recapture experiments
and growth data were derived for each age group by length
frequency and scale analyses. The food consumption for
the population was calculated by summing the 4 h food
consumption results and then combining these daily ration
estimates with biomass data.

Results

Locomotor and Feeding Rhythms

Locomotor activity All age groups showed diel rhythms of
locomotor activity, although these changed considerably
throughout the year. Young fish were mainly diurnal with
respect to their locomotor activity throughout the summer
and autumn (January - May), as compared with the more
nocturnal* behaviour of older fish (Fig. 1). In winter
(July), however, all fish became diurnal. The spring period
(September - November) was characterised firstly by older
fish reverting back to being nocturnal, followed by a
period during which all age groups were extremely active
over both the day and night. This extended period of activity resulted in an overall activity level for this month
greater than that recorded for any other month (Fig. 2).
Thus, although the seasonal cycle of locomotor activity
was roughly correlated with annual temperature changes,
the maximum activity did not coincide exactly with maximum water temperature.

Associated with the diel change in overall locomotor
activity, a significant onshore-offshore movement occurred
among the various age groups (Fig. 3). In summer, young
fish frequented the shallow water around the edge of the
lake by day, but were replaced by older fish at night.
In winter, all aged fish were distributed at random over
the lake by day, but young fish were more numerous around
the edge at night (G values 15.47 - 78.53, d.f. = 4;
$P < 0.01$).

Feeding activity In general, periods of higher feeding
activity were associated with periods of higher locomotor
activity. This is shown firstly by the seasonal change
in the feeding pattern of the age 1+ fish (Fig. 4).
During autumn, summer and winter, both feeding and movement were higher during the day than during the night;
in spring, at a time when all fish became active during
the day and night, feeding also extended to both these

---

*Footnote: Because nets were set for the whole night period, crep-
uscular behaviour could not be distinguished from true nocturnal
behaviour.

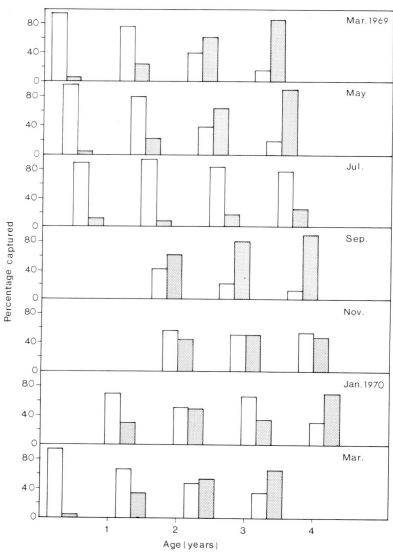

Fig. 1. Seasonal changes in the diel locomotor rhythms within each age group (sexes combined). Diel activity is given as the average catch per hour of fish caught in trap nets during the day and during the night expressed as a percentage. (Redrawn from Staples, 1975a.)

periods. Secondly, during the summer when young fish were day-active and older fish were night-active their feeding activities showed a similar pattern (Fig. 5). Fig. 5 also

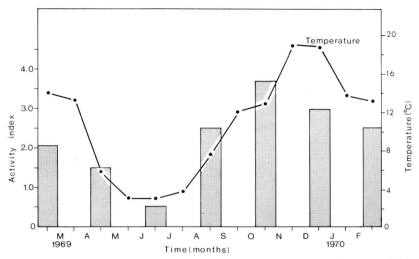

Fig. 2. Seasonal variation in locomotor activity level (shaded histogram) and the mean water temperature from March 1969 to March 1970. (Redrawn from Staples, 1975a.)

suggests that although locomotor and feeding activities were associated, the peaks of the two activities did not coincide exactly. For example, the locomotor activity of age 1+ and age 2+ fish was most intense just after dawn, but feeding activity did not reach a peak until several hours later. The sequence suggests that increase in light at dawn triggered movement in these fish, and then, while activity was high, the fish began feeding.

Thus, as a result of the complex seasonal phase shifts, all age groups showed similar diel periodicity in both locomotor activity and feeding during the winter and spring. Adult fish, however, changed from being day-active in winter to being night-active in summer and, consequently, in summer and autumn there was little overlap in the periods of high activity of young and old fish.

Production Dynamics

The total annual production (including the production of sexual products) during the year of study, February 1969 to February 1970, amounted to 39.8 g/m² and the total food consumption was 251.9 g/m². A marked seasonal cycle of production was recorded which was influenced by changes in both the amount of food consumed and the efficiency with which the food was channelled into growth (Table 1). Approximately 70% of the annual production occurred in the late spring - early summer period from November to January, at a time when both food consumption

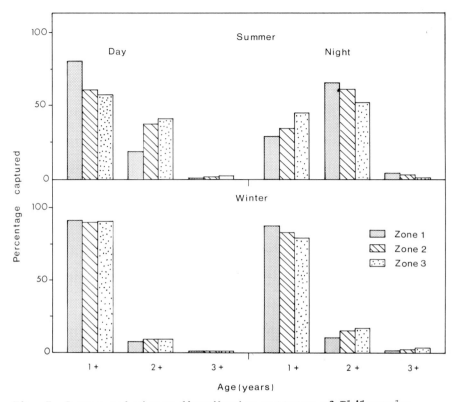

Fig. 3. Summer and winter distribution patterns of *Philypnodon breviceps* in the lake expressed as a percentage composition of age groups in each concentric zone. (Zone 1, shore - 10 m; zone 2, 10-25 m; zone 3, 25 m - centre). (Redrawn from Staples, 1975b.)

and gross growth efficiency were high. In contrast to the spring period, growth efficiency was very low in autumn (although water temperatures were similar) and despite a relatively high food consumption and high bio-mass, growth and production were low. By winter, food consumption had also declined markedly and little pro-duction resulted.

Discussion

Staples (1975a) showed that food passage rate and hence food consumption of *Philypnodon breviceps* was dependent on both the prevailing water temperature and the amount of food present in the gut. Diel feeding rhythms, there-fore, could play an important part in controlling food intake and hence directly affect the growth of these fish. For most of the year, feeding in all age groups was limited

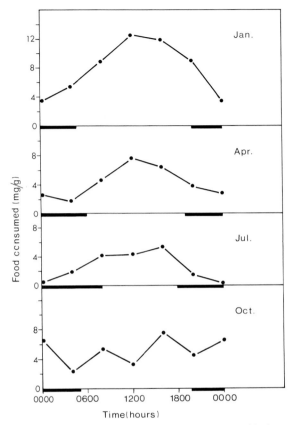

Fig. 4. Seasonal changes in the diel feeding periodicity of one year-class (age 2+ fish) of *P. breviceps*. Shaded areas indicate hours of darkness. (Redrawn from Staples, 1975a.)

to one short period during the day or night. It was only during the spring period that feeding became extended to include both a day and night component and this resulted in a food intake higher than expected from temperature changes alone. The increased food intake appeared to stimulate rapid growth and production from November to January and an increase in the total biomass of fish present in the lake followed, reaching a maximum at the end of summer.

It was during the time of high biomass that asynchrony of age group behaviour was observed; young fish moved and fed during the day, while older fish became more active at night. Several other mechanisms which were observed to reduce intraspecific contact were also more pronounced during this time of the year. Young-of-the-year fish (age 0+) were largely pelagic during this period. The remainder

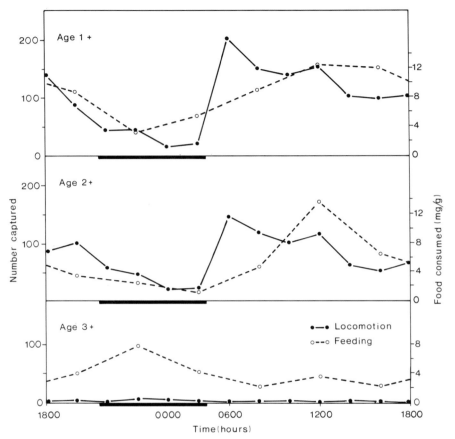

Fig. 5. Diel locomotor activity rhythms in January 1970 as measured by the number of fish of each age group (sexes combined) captured in trap nets during 2 h intervals (solid lines) shown together with diel feeding rhythms for the same month. (Redrawn from Staples, 1975a.)

of the population were benthic but exhibited a diel periodicity in onshore-offshore movement, older fish replacing younger fish around the shore at night. Fish of different ages were also selecting different prey, the difference being most marked at this time of the year when older fish were feeding at night. All the differences in behaviour (both in space and time) would have the effect of segregating age groups into separate 'niches' within the population, thereby reducing competition between these groups. That this had in fact occurred was demonstrated by the negative correlation observed between numbers and growth within a particular age group (Table 2). Mainly as a result of an almost complete lack of spawning during 1969,

TABLE 1

Seasonal changes in the total population number ($\hat{N}$ x 10^3), mean biomass ($\bar{B}$ (Kg)), production (P (Kg/month)), food consumption (C (Kg/month)) and gross efficiency of food conversion (K = P/C).

	Autumn (Mar. - May)	Winter (Jun. - Aug.)	Spring (Sep. - Nov.)	Summer (Dec. - Feb.)
$\hat{N}$	1083.00	291.00	179.50	161.90
$\bar{B}$	143.88	82.70	95.93	182.93
P	8.94	3.81	22.72	50.83
C	252.60	55.49	116.56	225.37
K	0.035	0.069	0.195	0.226

TABLE 2

Comparison of density (No./m^2), mean body weight (g) and biomass (g/m^2) of each age group between two consecutive summers (February 1969 and February 1970).

Age Group	February 1969			February 1970		
	Density	Mean Weight	Biomass	Density	Mean Weight	Biomass
0+	171.0	0.047	8.04	0.2	0.412	0.08
1+	33.3	0.430	14.32	12.5	1.211	15.14
2+	2.9	1.750	5.08	11.5	1.652	19.00
3+	0.3	3.301	0.99	0.5	3.177	1.59
Total	207.5		28.43	24.7		35.81

there was a very small number of age 0+ fish present in February 1970. The mean weight of this particularly weak year-class, however, was almost ten times that reached by the age 0+ fish at the same time of the preceding year. Similarly, the number of age 1+ fish present in 1970 was considerably fewer than in the previous year, but this was compensated for by increased growth such that the biomass of this age group at the end of summer in both years was essentially the same. The two older age groups were more numerous than their counterparts a year earlier and were compensated for by correspondingly lowered growth.

Hence, in terms of the total population biomass, the

effects of year-to-year variability in year-class numbers were greatly dampened. Thus, the total biomass of the population was strikingly similar in two consecutive summers despite large differences in the population density and age structure. The population density recorded in February 1970 was only 12% of that of the previous year, yet the biomass of 35.81 g/m² was even slightly greater than the 28.43 g/m² measured in February 1969.

In summary, the presence of diel rhythms of locomotor and feeding activities appeared to influence the production dynamics of P. breviceps in two ways. Firstly, there was a direct effect on the growth rate of the fish imposed by having only a restricted feeding period each day for most of the year. Secondly, there was the indirect effect of growth compensation within age groups which resulted, in part, from the asynchronous behaviour of these groups. Obviously, other mechanisms which tended to segregate age groups in space as well as time were also affecting the level of intraspecific competition. However, diel rhythms were extremely variable throughout the year and it appears that it is this flexibility which amplifies their significance. For example, the change in diel feeding pattern to an extended period encompassing both the day and night occurred at a time of rapid elaboration of both body and gonad tissue. Then, as the population biomass increased, the age groups became asynchronous in their behaviour, a factor probably contributing to the reduction of competition among age groups at this time. This reduced competition was manifested in a growth compensation phenomenon acting within rather than between age groups, such that a density-dependent regulatory mechanism was set up tending towards stability in annual production and resulting biomass.

References

Allen, K.R. (1969). In: Symposium on salmon and trout in streams (T.G. Northcote, ed.), pp.3-18. Institute of Fisheries, University of British Columbia, Vancouver.
Backiel, T. and LeCren, E.D. (1967). In: The Biological Basis of Freshwater Fish Production (S.D. Gerking, ed.), pp.261-293. Blackwell, Oxford.
Crowe, W.R. (1950). Progve Fish Cult. 12, 185-192.
Hopkins, C.L. and McDowall, R.M. (1970). Proceedings Part I. Background statement of the New Zealand water conference 1970.
Ricker, W.E. (1946). Ecol. Monogr. 16, 375-389.
Smyly, W.J.P. (1957). Proc. Zool. Soc. Lond. 128, 431-453.
Staples, D.J. (1975a). J. Fish Biol. 7, 1-24.
Staples, D.J. (1975b). J. Fish Biol. 7, 25-45.
Staples, D.J. (1975c). J. Fish Biol. 7, 47-69.
Stott, B. (1970). J. Fish Biol. 2, 15-22.
Strawn, K. (1954). Copeia 195-197.

A REVIEW OF SONAR TECHNIQUES FOR COUNTING FISH*

H. BRAITHWAITE

Vickers Oceanics Ltd.,
Barrow-in-Furness, Cumbria, England

Introduction

The basic idea of using sound waves to count fish migrating up a river system is attractive in that a directional beam of sound energy can propagate over long distances, so the system can be used over large river sections without the need for any form of barrier across the river. There are two basic approaches to the problem; one is to send out a pulse of acoustic energy and receive the reflected echo signal from a fish and hence determine its presence. The alternative is to project an acoustic beam across the river and receive the energy at the opposite bank. If an object like a fish is in the line of sight between the transmitter and receiver, the receiver energy will be reduced and the presence of the object detected.

The directional sound waves used must be produced from reasonably small transmitters (transducers) and to achieve this, acoustic theory dictates that the frequency of the sound waves must be in the ultrasonic region, i.e. hundreds of kilohertz. However, the attenuation of the sound waves increases as the square of the frequency so the upper limit of the usable frequencies is determined by the range required. The engineering compromise puts the operating frequencies in the range of 0.5 to 1 MHz.

Principles of Sonar Counters

Echo Type

One of the earliest counters using acoustic techniques was the Bendix Counter (Anon., 1966) which consists of an

---

*Footnote: Work carried out when the author was with the Dept. of Electronic and Electrical Engineering, University of Birmingham P O Box 363, Birmingham 15.

array of transducers laid across the river bed and looking
to the surface, as shown in Fig. 1.

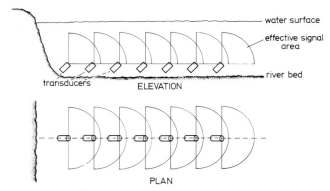

Fig. 1. Arrangement of Bendix Counter

The transducers are sequentially pulsed and fish above the
array give an echo which is processed to give a count.
This is a statistical counter since a single fish remaining in the beams is likely to be counted several times.
Also, since there are no directional properties, debris
floating downstream, if it gives a sufficiently large
echo, is processed as a fish. The system is only really
usable in rivers with large numbers of upstream fish,
very little debris and at least 1m of water depth.
    The alternative approach developed at Birmingham Uni-
versity is to place two acoustic curtains across the river
separated by a small distance, and to observe the order in
which the fish pass through the beams (Braithwaite, 1971,
1974a). The beam arrangement is shown in Fig. 2.

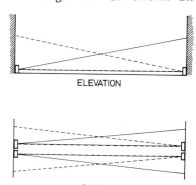

Fig. 2. Transducer beam arrangement of counter.

The frequency of operation is 1 MHz. The electronics of

the unit process the echoes to determine the direction of
the fish and register them on an electromechanical counter.
The development of the above system has been to try and
obtain unique echo signatures from fish and to this end
processing has been evolved which compares the echo signals
from successive pulse scans. The difference output is zero
for stationary targets, but for a moving fish the echo
amplitude varies from pulse to pulse so the difference
unit gives an output. This moving target detector which
uses the same principle as in radar (Scholnik, 1962), dis-
criminates between echoes from moving fish and rocks on
the river bed. Although this technique does discriminate
fish from stationary targets there are site problems which
produce inanimate moving targets as will be discussed
later.

The Interrupt Type Counter

If an acoustic beam is transmitted across a river with
a receiver on the opposite bank, then the presence of an
object like a fish in the line of sight between the transducers will produce a reduction in the received energy.
This reduction in energy can be used as a means of detecting the presence of the object. This principle has not
been seriously used because in the early days the idea of
a sonar counter was for full depth coverage to be achieved,
which would entail a number of transmitters and receivers
to give full line of sight coverage of the river depth.
However, research has shown that the fish tend to swim
near the river bed in fast flowing water so one or two
transmitter/receivers would suffice to cover the relevant
bottom section of the river. The chief advantage of the
system is that averaging can be used on the received signals which would eliminate electrical interference spikes.
The system would then become very similar to the conductivity counter without the need for electrodes across the
river bed.

Doppler Sonar Counters

The principle of a Doppler system is that echoes from
stationary objects are received at the same frequency as
that transmitted, whereas echoes from moving objects are
received at a slightly different frequency, the difference
being proportional to the velocity of movement along the
beam axis. Velocities perpendicular to the beam axis give
no Doppler shift. Unlike the echo amplitude method of
detection a Doppler system can distinguish between echoes
from fish and rocks by classifying them according to frequency.

There are two basic approaches to the use of Doppler,
one is to measure the Doppler shift of the translational
movement of the fish and the other is to measure the frequency shift caused by the tail and body oscillations.

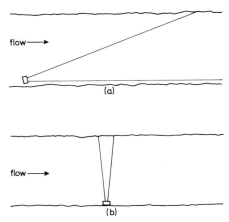

Fig. 3. Siting of sonar beams for Doppler systems: (*a*) to optimise translational Doppler (*b*) to optimise body movement Doppler.

The siting of the sonar beams to optimize the two effects is different, as shown in Fig. 3. The first approach has the disadvantage that a large volume of water is insonified, which increases the unwanted background noise and this, combined with the unpredictable upstream velocity of the fish and the low target strength of the fish in the head-on aspect, makes the system unattractive and not a practical proposition. The second approach is the one that offers the best possibility of a real solution. The frequency shifts for a 500 kHz transmission are in the region 0-4 kHz, i.e. audio-range (Babb, 1974). In practice the performance of such a system depends both on the magnitude and frequency of fish tail movement and on flow conditions in the river. The acoustic signal transmitted will be scattered back towards the receiver by air bubbles suspended sediment and vegetation, and under turbulent flo conditions this backscattered signal may suffer an appreci able frequency shift. Doppler signals from the fish tail will only be separable from these 'noise' signals if both the amplitude and frequency shifts are large enough.

There are two different principles used in Doppler sonar systems, one uses a continuous wave transmission, which gives infinite frequency resolution but no range information. The alternative is a pulsed system where the Doppler resolution is related to the pulse length transmitted and the system is much more complex. Most of the work in rivers to date has been done with the simpler continuous wave system and to reduce the reverberation at short ranges a bistatic system, as shown in Fig. 4, can be used. This can give an improvement in the signal to noise ratio of 30 dB's.

Work done to date on the fish body movement system has

shown that an automatic system of detecting the Doppler
signals from the fish is not a simple task but the human
ear was able to identify the 'swooshes' of the fish tail.

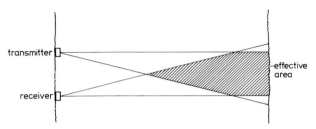

Fig. 4. Bistatic beam configuration.

This has been used as a means of checking, by siting the
system upstream of a counter. The audio Doppler frequency
shifts are fed to a small cassette tape recorder and the
recorder is controlled from the counter and run for 20
secs after a count has been registered. Replaying the tape
at a later date, the presence of the fish Doppler signals
can be verified to confirm that the count was genuine.

Problems and Limitations of Sonar Counters

The ideal system of counting fish is to achieve a unique
signal from the fish. The amplitude of the echo gives no
indication of the form of the target, but using the moving
target detector the echo system can be made to distinguish
between rocks and moving targets. Unfortunately fish are
not the only moving targets and the big problems are reflections from surface irregularities, and multipath reflections via the water surface which are also moving.
The received echoes are short in duration and induced signals from electrical interference can cause problems.
The detection principle of the interrupt counter is to
measure the reduction of the received signal and it is not
possible to determine whether this was due to a single
large fish or a large number of small fish. Also multipath
via the surface may cause problems if the receiver beam-
widths are not kept small. This system will never be able
to give a signal unique to a fish.
The Doppler principle does give a signal which is unique
to a fish but the only discriminator at the moment is the
human ear because of the poor signal to noise ratio. This
ratio depends upon the amount of suspended sediment, flow
conditions, and the surface multipaths. The surface multipaths, because they have the moving surface as part of
the path, are continually varying and interference with
the direct path causes amplitude variations which result
in frequency modulation. The use of advanced signal processing techniques should eventually lead to an automatic
system for detecting the fish.

The Siting of Counters

Using a wide channel counter presents many problems associated with fish behaviour, in that unless the site is carefully chosen the fish will not always swim straight across the counting area, but may swim round and round or oscillate up and down the channel. Also, a resident fish population may swim around the counting area. This means the counter must be sited in an area of fast flowing water preferably in an unsheltered position. The severe attenuation of acoustic waves in aerated water (Braithwaite, 1974b) means that the site must not be close to turbulent white water or waterfalls. The problems of sharp surface irregularities limit the site to an area of smooth laminar flow.

Conclusions

The use of acoustic techniques in fish counting still requires more development before a reliable and robust form of automatic counter can be produced. The advancement of signal processing techniques and the availability of more complex integrated circuits, especially microprocessors, will eventually make the unique identification of a fish echo signal a possibility at an economic price. This will be some form of moving target detector or extractor of fish tail-beat Doppler signals. The siting of counters will always require careful consideration but the more advanced the signal processing system the greater the choice of site. The sonar counter will eventually be a more versatile counter than the conductivity type but will be a more complex system. Although the electronics unit will be expensive this will be more than offset by the low cost of installation.

References

Anon. (1966). Field Test Report, Sonar System for Salmon Counting. Bendix Corporation.

Babb, R.J. (1974). Fish Counting Using Doppler Sonar Techniques. Interim Report to Water Resources Board, from University of Birmingham.

Braithwaite, H. (1971). *J. Fish, Biol* **3**, 73-82.

Braithwaite, H. (1974a). *E.I.F.A.C. Tech Paper* **23**, *Supplement* **1**, 369-377.

Braithwaite, H. (1974b). *J. Sound. Vibr.*, **37**, 557-562.

Scholnik. (1962). *Principles of Radar*. McGraw Hill, New York.

ELECTRICAL RESISTIVITY FISH COUNTERS

D. SIMPSON

North of Scotland Hydro-electric Board Research Laboratory
Pitlochry, Scotland

Introduction

This paper presents an attempt to describe the basis and behaviour of electrical resistivity (or conductivity) counters in terms which it is hoped may be meaningful to potential users. The principles upon which their operation depends are described in no more detail than is necessary to appreciate the limitations upon practical performance.

Two types of installation are described:

(1) Tube counters which have been in operation in fish ladders in the Scottish Highlands since 1952;

(2) Strip counters which have been the subject of recent and continuing research. Although the original research specification has perhaps now been met some further fundamental information is still essential. There is also a requirement for an instrument with considerably enhanced capabilities.

Tube Counters

Introduction

Counting fish by noting changes in the electrical resistivity of water through which they swim was pioneered by Lethlean (1953). The basis of his approach was to apply a low voltage across two volumes of water through which the fish to be counted are constrained to pass and detect the changes in their resistances caused by the presence of the fish.

Method

Lethlean's original counters were located in the submerged orifices connecting the pools of the fish ladders which the North of Scotland Hydro-electric Board were building around their newly constructed dams. They took the form of 3 electrodes approximately 1 fish length apart around the internal periphery of an electrically

insulated tunnel. They were energised at a much lower A.C.
voltage than could be felt by a large fish and the water
resistance between each pair formed part of a Wheatstone
bridge (Fig. 1) which was completed by fixed resistors
within its electronic detecting apparatus.

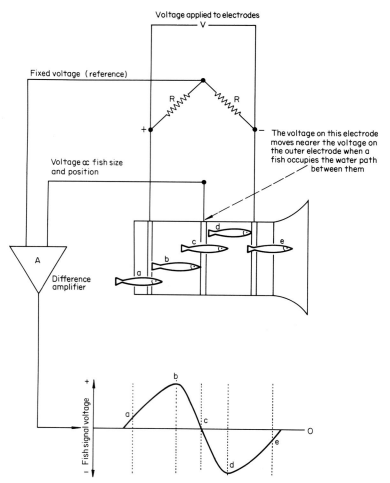

Fig. 1. Wheatstone Bridge arrangement of Lethlean fish counter.

This device, designed by A.M. Clark of the North of Scotland Hydro-electric Board, amplifies the 'out of balance'
voltage caused by the presence of fish, passes this signal through logic circuitry and activates an appropriately
labelled digit meter. The direction of travel of the fish
is known from the order in which the signal changes occur

and a rough measure of the length of the fish can be obtained from the peak signal amplitudes. The counter in-
corporates one or more adjustable length discrimination
settings.

Limitations

Any living creature of comparable or larger size than
the prescribed minimum length of fish to be counted, or
indeed any object causing a sufficiently large change in
the electrical resistivity of the water, will be registered. However, the strength of the resistivity method
is that the number of false counts due to non-fish signals
appears to be less than would arise with other methods of
detection. Its most apparent weaknesses are listed below.

The counter can only discriminate between fish of different species on a measure of their length. The accuracy
of any count is therefore partially dependent upon the
counter's ability to distinguish between fish of different
lengths and in this respect there are two unavoidable
sources of inaccuracy:

    (1) Since the electrical field is weaker in the
    centre of the tube than nearer the electrodes the
    same fish will cause a smaller signal if it swims
    through the centre of the tube than it would do if
    it hugged the walls;
    (2) Due to its undulating swimming motion, the fish
    may not be fully stretched when it occupies the
    position where it gives the largest (peak) signal.

When setting discrimination levels an attempt is made
to arrange that the numbers of small fish included erron-
eously is balanced by a like number of large fish which
are excluded. This of course means that if two different
species are to be counted separately then the accuracy by
which they are separated varies with the amount of overlap that exists in their length ranges.

A further source of counting error would arise if two
or more fish were to pass through the tube together. Since
the counter can only detect changes in electrical resistance these fish would merely register as one (slightly
larger) fish.

Avoidable Errors

If fish were to linger between either pair of electrodes
thereby holding the bridge out of balance then any other
fish swimming through would go undetected. For this reason
the water velocity in the tube should be kept sufficiently
high and this can be done most easily by ensuring that
there is a difference of at least 0.5 metres between the
water levels upstream and downstream of the tube.

Finally if the water level upstream of the tube falls
too low, then a large vortex may form and appreciable
volumes of air may be drawn down through the tube. Since

the electrical resistivity of air differs from water in
the opposite sense to that of fish, the signals caused
by air moving downstream are similar to those due to a
fish swimming upstream. The formation of such a vortex
could therefore create many false counts. To avoid this
the water level upstream of the counter should be maintained at least 0.5 metres higher than the soffit or top
of the tube.

Efficiency

   Although specific measurements have not - to my knowledge - been made, it is expected that a well sited tube
counter in a well designed fish ladder will count with
an efficiency of better than 95%.

Strip Counters

Introduction

   In 1968 the Natural Environment Research Council (NERC)
set up a working group to 'review the needs, guide the
devising and development and advise on the scale, timing
and funding of research into automatic fish counting'.
The Group set their initial target as an 80% or better
count of ascending fish only, with no restriction allow-
able to the width of the river. Lethlean's electrical resistivity method might successfully and economically detect fish as they swam up the face of the Crump type flow
measuring weirs being constructed by the English River
Authorities. To apply Lethlean's method without the tubes
it was necessary to find a flow situation with a water
surface as still as possible since unlike the tube, fluctuations of the water surface create 'background' signals
which must be kept appreciably lower than those from the
smallest fish to be detected. Previous attempts to achieve
this had sited electrodes in relatively still water. The
innovation here was that they were deliberately placed in
the region downstream of the weir where the water flow
had its maximum velocity (Fig. 2). Apart from the still
glass-like surface relatively undisturbed by the wind
this region had two further advantages:
   (1) The high velocity would ensure that no fish
   lingered over the electrodes to keep the counter
   out-of-balance;
   (2) The lower velocities near the bed should per-
   suade the fish swimming upstream to do so near the
   electrodes thereby causing a larger signal.

Method (See also Simpson et al., 1975)

   In effect Lethlean's strip electrodes were unwound
from the tube and laid along the downstream face of the
shallow weir. The method should be equally effective in
any situation where this rapid glassy surfaced flow arises
and the Freshwater Biological Association (FBA) have been

Fig. 2. Strip counter electrode in position under smooth water surface, downstream of weir.

using it with some success since they moved their electrodes into the throat of their Venturi Flume (the narrow portion of a restriction in the river width) on the River Frome in 1969.

Clark re-examined the details of the design of his counter to achieve maximum amplification and stability and added an automatic bridge-balancing device which made slow adjustments to the bridge to preserve its balance against the slow changes in water surface profile. The device also comes into play when fish swim over the electrodes but they do this so much more quickly that the amount of adjustment is then negligible and does not interfere with the counting arrangements.

Apart from this there was no change from the counter used with tubes and all of the research effort was spent upon learning how long the strip electrodes could be before the signal from a fish became undetectable. The longer each set of electrodes the smaller is the number of electronic counters required to cover the full width

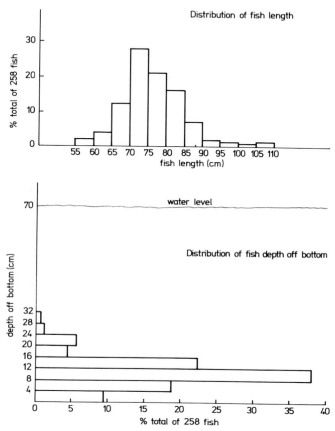

Fig. 3. (*a*) Distribution of depth of fish swimming over counter at East Stoke. (*b*) Distribution of body length of fish counted at East Stoke.

of the river and therefore, the cheaper is the installation at any particular site. The limiting factor is the ratio of the strength of signal from the fish (which becomes smaller as the electrodes are extended) to the continuous background signal coming mainly from the water. This explains why increase in the 'water signal' necessitates a similar proportional increase in the minimum detectable fish signal and therefore a consequent decrease in the length of electrodes over which any given fish can be detected.

Limitations

All of those listed for tube counters still apply.
In addition, the extension of the electrode lengths to
their maxima reduces the probability of detecting fish

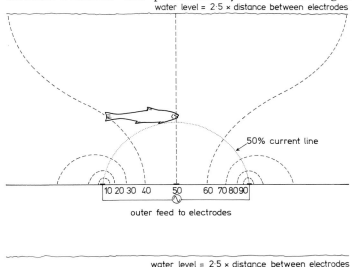

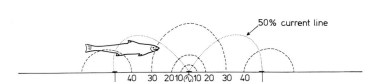

Fig. 4. Arrangement of feed to electrodes at East Stoke. Upper
diagram - original system. Lower diagram - present system.

moving downstream. The reasons for this are that such
fish have no need to swim low in search of low velocities
and sometimes even drift downstream obliquely thereby
further reducing the strength of the resulting signal. If,
as in tube counters, the signal strength is sufficiently
high then these reductions do not cause concern but if
the signal is decreased to its minimum for economic reasons, then a conflict of interest arises. This conflict

was accentuated when tests at the FBA's River Laboratory,
East Stoke (Figs. 3 and 3A) (Hellawell, 1973) confirmed
that ascending fish did indeed swim low and it became
possible to change the method of applying the electric
field to the water. This was done to decrease the strength
of the electric field at the water surface thereby decreasing the 'water signal' (Fig. 4). Smaller fish signals
could then be detected and the length of weir over which
ascending fish could be counted was extended. The variation of signal strength with fish swimming depth was, of
course, increased, consequently making it even more difficult to detect descending fish (Fig. 5).

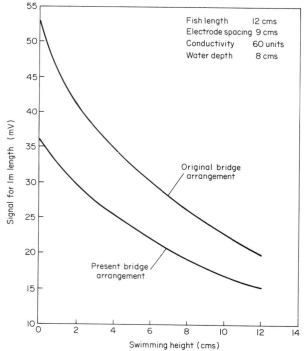

Fig. 5. Relationships between signal strength and distance of fish
above counter.

Avoidable Errors

The problems of fish lingering between electrodes is
unlikely to arise here because of the high water velocity
but low water levels upstream must still be avoided. If
the depth of water over the electrodes is too small then
the small variations at the surface are detected by the
counter and many false counts again arise.

This flow situation is undesirable for other reasons

and it is suggested that when weirs are built a low section should be left of such a width that the water level
upstream is at least 0.2 metres higher than the crest of
this section of the weir at the lowest river flow. In
addition to avoiding fish counting problems, this will
also ensure that fish have sufficient swimming depth of
water over the weir at all times and that water flow
measurements are not made inaccurate due to weir surface
problems.

Algal growths (which have lower electrical resistivity
than water) over the counting zone can create spurious
signals and must be removed before they become too extensive.

A final source of error may arise if the downstream
water level is allowed to rise so high that the standing
wave (white water at the downstream end of the glassy
surface) moves up over the electrodes.

This can be avoided by arranging for a clear water
channel downstream.

Efficiency

An independent assessment of the counter's efficiency
was made on a 15 metre long crump weir with electrodes
spaced 457 mm apart during the 1976 salmon run with the
following results (from Beach, 1977):

'8.2 *Conductivity Counter*
The conductivity counter gave a high correlation
between recorded counts and visual observations
with a maximum correlation coefficient of 0.977
being obtained. A quantative test of the counter
accuracy requires that the minimum length of fish
detected be determined and for the small number of
day-time fish (132) it was found that of the 41
which were longer than just under 18 ins. (457 mm),
all were counted'.

Acknowledgements

Without financial support from the North of Scotland
Hydro-electric Board, the Natural Environment Research
Council, the Water Research Centre, the Department of
Agriculture and Fisheries for Scotland and the Water Data
Unit and the co-operation of the Freshwater Biological
Association and many River Authorities, the work reported
above would not have been possible. For this and their
permission to present this paper I am grateful. I would
also like to thank my colleagues in the Pitlochry Laboratory for their assistance and fellow-workers elesewhere
for their co-operation.

References

Beach, M.H. (1977). *Determination of the accuracy of two types of
fish counter, employing the principles of: i) conductivity change,*

and ii) acoustic reflection. Report of Manley Hall Project,
  M.A.F.F., London. 27pp.
Hellawell, J.M. (1973). *In: International Atlantic Salmon Symposium*
  (M.W. Smith and W.M. Carter, eds.), *Int. Atl. Salm. Found. Spec.*
  Publ. **4** (1) 317-337.
Leathlean, N.G. (1953). *Trans. Roy. Soc. Edinb.* **62**, 479-526.
Simpson, D., Clark, A.M., Slessor, M.D., and Dudgeon, I.D. (1975).
  In: Symposium on the Methodology for the Survey, Monitoring and
  Appraisal of Fishery Resources in Lakes and Large Rivers. Aviemore
  1974. (R. Welcomme, ed.), E.I.F.A.C. Tech. Pap. 23, Suppl. 1,
  414-435.

A REVIEW OF BIOTELEMETRY TECHNIQUES USING ACOUSTIC TAGS

R.B. MITSON

*MAFF, Fisheries Laboratory, Lowestoft,
Suffolk, UK*

Introduction

Fish tracking in the present sense is a relatively new technique, the first experiment being recorded about 20 years ago (Trefethen, 1956). Long before this acoustic transmitters had been in use for position fixing in the sea, but the thermionic valve technology did not lend itself to the attachment of such transmitters to fish. The use of miniature electronics became possible with the advent of the transistor but there was a period when progress was slow whilst other small components and batteries were developed.

The level of electronic technology currently being applied to the design of acoustic fish tags is below that of the 'state of the art' although it is able to satisfy many of the present requirements. As tracking and telemetry develop higher levels of technology must be used so that complex circuits can be accommodated within the volume taken by the basic circuits of today. The necessary degree of micro-miniaturization will require the use of expensive facilities which may be difficult to justify unless there is some attempt to standardize systems. A review of current techniques may serve to show the extent to which the requirements are being met at present and to give an indication of possible extension in the future.

Before making this review some comment on the terminology used for tracking and telemetry may be worthwhile for there is a tendency to isolate the two words. Bio means life and telemetry is the remote indication of information, so in combination these words offer a suitable compromise to cover fish position fixing and the monitoring of physiological and physical variables from it and its immediate surroundings. A collective description of the techniques referred to in this paper might be underwater acoustic biotelemetry.

Applications

Studies of the behaviour of fish are difficult in rivers, lakes and seas because continuous observation is rarely possible. As a result progress was slow before acoustic methods were available and the possibilities of these are still being explored. In principle these methods of fish monitoring allow the study of individuals and their activity to be related directly or indirectly to the environment. The techniques rely on tagging the fish, i.e., fitting a transmitter to it which transmits through the medium of the water to a remote position, or a number of positions, where the observer's receiving equipment is situated. Stasko (1971) has reviewed field studies on fish orientation. Biotelemetry in its simplest form requires the description of the movements of individual fish, which can involve a scale of activity monitoring from within a few cubic metres in a laboratory tank or in a river to many kilometres in the sea. By surveying the environment in which the fish is living, or by having previous knowledge of the area, it may be possible to relate activity to one of the physical variables, current speed or direction, temperature, oxygen, salinity or to the presence of a pollutant. Environmental conditions are often complex and fast changing. So to establish a direct relationship between these and fish activity it may be necessary, not only to sense and telemeter the physical variables of the medium surrounding the fish, but also the physiological changes taking place within it, all from sensors attached to the body. This is an ultimate goal, only technically feasible at present for very large fish, because miniature sensors are not yet developed for all the functions or variables to be measured. But there is reason to suppose that these will be forthcoming in the next decade.

At present techniques are required to give momentary, hourly, daily, weekly, or monthly patterns of activity to fit the aims and scale of particular experiments. If the location of a fish is fixed vertically by its depth in the water column and horizontally with respect to known geographic axes, or other points of reference, its detailed movements can be ascertained within the limits set by the positional accuracy of the techniques and the rate at which observations are made.

Accuracy of Location

When fish are being tracked it may be sufficient to detect gross movements over the period of study. But in detailed studies of activity and behaviour the position of the acoustic tag must be known within well defined limits of accuracy. Because the acoustic beam of a directional hydrophone widens as the range from it increases, the greater the range between an acoustic tag and its receiving hydrophone, the lower is the accuracy within which the position can be fixed.

To estimate an overall positional accuracy the parameters of the shipboard navigation system or reference stations must be known in addition to the geometry of the acoustic beam and the acoustic propagation conditions. Vessel navigation systems used in biotelemetry range from visual sights of landmarks on rivers or lakes to the Decca Navigator system in the open sea. Geometry of the acoustic system takes into account the receiving beam dimensions, how it can be steered or manipulated; or the disposition of bottom laid hydrophones. The transmission from the tag plays no direct part except to provide a sufficiently high level of signal to prevent ambiguous bearings being taken where the signal to noise ratio is low.

The simplest tracking system uses a directional hand steered hydrophone in conjunction with a 'pinging' acoustic tag, i.e., it transmits a continuous train of pulses or 'pings'. A receiving hydrophone of small physical dimensions for easy handling has by definition a relatively wide beam. With this simple system the range is not known, thus the tag signal could originate anywhere within the volume swept out by the beam angle as range from the hydrophone increases (Fig. 1).

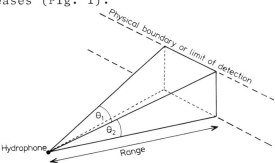

Fig. 1. Diagram of beam of simple directional hydrophone.

The limit of the range R is set by the distance to a physical boundary across the axis of the beam, or the maximum detection range of the equipment. The signal will be somewhere within the volume for a rectangular beam.

$$V \ (m^3) \ = \ \frac{4R^2}{3} \ (Cos\theta_1 . Cos\theta_2)$$

where: R is the range in metres as defined above

θ_1 is the full beam angle in the horizontal plane
θ_2 is the full beam angle in the vertical plane

e.g. if $\theta_1 = 5^0$, $\theta_2 = 7.5^0$ and R = 150 m, V = 29628 m³.

Rectangular faced hydrophones are necessary if differ-

ent beam angles are needed in the two planes. A long face
in the vertical plane gives a narrow beam to reduce surface and bottom reflections and in the horizontal plane it
will give better bearing resolution. These must be constructed of a number of elements to obtain narrow enough
beams for reasonable bearing resolution and so they tend
to be expensive.

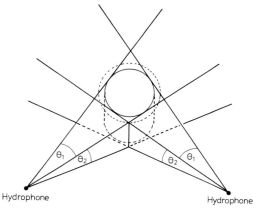

Fig. 2. Cylinder of uncertainty of position of a tag, determined by
triangulation with simple directional hydrophones.

In practice the situation would only be as bad as the
above example if the observer had no way of moving the
site of his hydrophone. Usually it can be moved by boat
or land transport and another bearing taken as illustrated
in Fig. 2. If the line between the two points from which
bearings are taken is used as a reference baseline, a
horizontal (plan) position of the tag can be calculated
by triangulation. An approximate range from each position
will enable the beam width at the tag position to be obtained and a circle of uncertainty drawn as in Fig. 2.
To this must be added an outer circle which allows for
the maximum distance that the fish could have moved between
the position fixes. Experienced operators can often estimate range by swinging the hydrophone beam at a constant
rate and noting the length of time a signal can be detected. In very shallow waters position fixing in the horizontal plane only may be sufficient, but if this is not the
case the circle of uncertainty becomes a cylinder whose
height is equal to the depth of water from surface to
bottom. If the hydrophone beam can be depressed in the
vertical plane the depth of a pinger is given by depth
(m) = R sin (ϕ + Θ_2) where ϕ is the angle of depression.
When R is an estimate and Θ_2 is wide only a very rough
estimate of depth can be made.

A fixed location system for use in a closed area of

water has been described by Young *et al.* (1972). Two or
more directional hydrophones are situated in the area in
such a way that their beams when mechanically steered can
be made to cross at the position of the tag signal. The
accuracy of position fixing is the same as for the system
having two hydrophone positions but with no uncertainty
due to fish movement between fixes.

To study the changes of depth by fish or to reduce the
uncertainty of fish position, tags have been developed to
measure hydrostatic pressure. Luke *et al.* (1973) have described a tag 70 mm long by 16 mm diam whose pulse repetition rate varies linearly with depth. The sensor is a
strain gauge which enables an accuracy of ± 0.35 m to be
maintained over the depth range of 0-40 m and temperature
variation of 5-20°C. This type of sensor requires a continuous current if its best characteristics are to be pre-
served. As a result a large battery is required and this
restricts the use of this sensor to large fish. The battery gives a life of 3 days for this tag.

An underwater adaptation of the hyperbolic navigation
system used by ships and aircraft has been used by Hawkins
et al. (1974) for fish tracking. Omni-directional hydrophones are laid on the seabed or suspended in midwater,
the distances between them being accurately measured. A
pinger is placed by each hydrophone in turn and the time
of arrival of its signal at each of the receiving points
is recorded. The difference in these times multiplied by
the speed of sound in water defines hyperbolae which intersect at the pinger position. For good results more than
3 hydrophones need to be laid but the 3 situated in the
best positions with respect to the fish tag at any particular time are used. Positional errors in such a system
are mainly due to local changes in the speed of sound
and a theoretical accuracy of about ± 1 m is possible in
the horizontal plane.

In situations where the fish is not confined to a given
area, but its detailed movements must be recorded, a transponding acoustic tag is needed. This tag responds to an
interrogation pulse from an echo ranging sonar by instantly
transmitting a pulse in return. Thus the time between the
transmission of the interrogation pulse and the reception
of the responding pulse at the sonar gives an accurate
measurement of range. A high rate of range sampling is
possible, limited only by the transit time of the interrogation and responding pulses. If the range between sonar
and transponder is 150 m the sample can be taken at 200
millisecond intervals. Because of the short transit times,
electronic timebases and displays are needed. The most
significant source of error in range measurement is due
to variation in the speed of sound because of temperature
and salinity changes. Table 1 shows typical variations;
a 5°C change giving rise to a speed difference of about

1%. The resulting range error is small compared to the uncertainty of position due to the width of the acoustic beam and lack of depth information.

TABLE 1

Speed of Sound. m/s

d = 5 m	Freshwater	Salinity %		
		30	32.5	35.0
t ^{0}C	Speed			
2.5	1413	1451	1454	1457
5.0	1424	1461	1464	1467
10.0	1445	1481	1484	1487
15.0	1464	1498	1501	1504
20.0	1480	1514	1516	1519
25.0	1496	1527	1529	1532

Fig. 3 shows that the position of a transponding tag would lie on an area bounded by the arcs subtended by the angles Θ_1 and Θ_2.

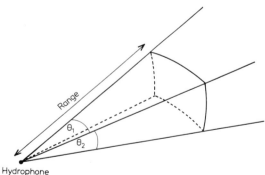

Hydrophone

Fig. 3. diagram of tag position (curved surface bounded by arcs subtended by the angles Θ_1 and Θ_2) determined with a transponding tag.

A sonar described by Voglis and Cook (1966) and Mitson and Cook (1971), when used in conjunction with a transponder (Mitson and Storeton West, 1971) allows detailed movements of fish to be observed in relation to surrounding objects (Harden Jones *et al.* 1977). The acoustic beam is very narrow (0.3°) in one plane which is scanned electronically over a 30° sector. The scanning beam can operate in either the horizontal or vertical planes so that

bearing and elevation (depth) fixes can be made on the
transponder.

Frequency

The small choice of carrier wave frequencies in commercial tags is limited mainly by economic reasons to 50 to
100 kHz with 74 kHz as the most common. However, it is
the technical requirements that must be considered for an
optimal design in a particular application. There are two
main design factors relating to frequency; the range of
detection and the size of transducer. The higher the frequency the shorter will be the range for a given power input; but the lower the frequency the larger the transducer
becomes. For a transducer operated at resonance only one
of the frequency dependent dimensions is used, and in the
case of a cylinder this can be either the diameter or the
length. If a low frequency is essential a length mode
resonance tube would offer the best results for low drag,
i.e., it is long and thin, but its radiation characteristics are more directional than those of a very short cylinder resonant in the circumferential mode. A short cylinder of large diameter lends itself most easily to internal
tags where the drag is not important. Fig. 4 shows the
diameter against frequency of transducer cylinders having
a height of 3 mm, and the deviation from omni-directional-
ity on their axes. Transducers are often the second largest
component of an acoustic tag in terms of dimensions or
possibly weight, although the actual volume may not be
great. Thus there is reason to choose the highest possible
frequency commensurate with the acoustic power that has
to be produced for the range of detection and duration of
life. Acoustic waves are normally assumed to spread spheri-
cally from their source, the energy dispersing over a
larger volume as the range increases. At a point R metres
from the source it will have decreased by 20 logR dB regardless of frequency. In shallow water or where there are
side boundaries this law of spreading is modified and the
measured loss may be much less due to the channelling
effect of the boundaries.

There is a frequency dependent loss due to chemical
absorption in sea water. This increases with frequency
and is the most serious limitation to the use of frequencies above 100 kHz in the sea. However, when tags are
fitted internally to fish there are losses due to the
flesh and masking of the transmission due to the swim-
bladder and backbone which get progressively worse as
frequency increases.

At frequencies below 100 kHz weather effects such as
wind, waves, rain and hail produce noise, limiting the
range. These effects can be highly variable and often
localized, making predictions of range in any area as
difficult as that of the weather itself. The slope of

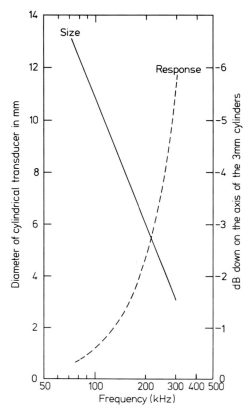

Fig. 4. Transducer size and response against frequency.

noise levels due to this type of effect falls at about
20 dB/decade reaching a minimum near 100 kHz. Other sources
of noise below 100 kHz are due to the engines of vessels
up to 1 kHz, but cavitation noise from propellors can extend to 50 kHz or higher.

At frequencies above 100 kHz a different source of
noise has to be considered, this is known as thermal noise
and is due to molecular movement in the water. The effect
on any receiving system can be calculated from the thermal
noise spectrum level:

$$N_t = -15 + 20 \log f - DI - E$$

where: f = frequency in kHz

DI is the directivity index of the hydrophone in dB
E is the efficiency of the hydrophone in dB.

Power and Duration

A long life is probably the most difficult design para-
meter for acoustic tags but biotelemetry must often extend
over long periods. Long life is not necessarily coupled
with a long range requirement; maximum duration of the
life of the tag has to be balanced against other parameters.

Where detailed movements of fish are to be studied, a
high rate of information is essential and this, coupled
with the maximum range and necessary duration of the work,
will set the battery specification. There are a limited
number of cell or battery sizes suitable as power sources
for acoustic tags. Two main types are used, mercury and
silver oxide, which are both available in similar size
and form. They have good voltage stability over their
working life but need to be stored at a low temperature
to maintain their capacity before use. Cells physically
small enough for this application have capacities ranging
from 12 mAh, with a volume of 80 mm^3 and weight of 0.28 g,
to 110 mAh, with 330 mm^3 volume and 2 g weight. These
cells are cylindrical in shape, and so convenient for
fitting in-line with the other components to produce a
low drag shape to the completed tag assembly. They can
account for up to 50% of the total volume and 40% of the
weight of a tag as shown in Fig. 5. These facts emphasise
the care needed in selecting the minimum sized battery
to give sufficient acoustic power for the duration of
the experiment.

Acoustic source level (SL) is the sound pressure to be
transmitted in order that a signal may be received with
a given ratio above the noise level (Signal to Noise Ratio,
SNR) at the maximum range of detection. The source level
may be read from fig. 1 of Mitson and Young (1975) if
the appropriate factors for directivity and sensitivity
of the receiving transducer and the receiver noise level
(limit of sensitivity) are applied. It is then possible
to calculate the power needed:

$$Pe_{watts} = \frac{1}{t\eta.c\eta} \text{ ANTILOG } \frac{SL-170.8}{10}$$

where: Pe = power in Watts from the battery

  $t\eta$ = transducer efficiency in %
  $c\eta$ = circuit efficiency in % and
  SL = source level in dB/μPa/m.

The power from cells at full rated capacity (mAh) is given
by the manufacturers at a specified current drain, usually
1 mA for small cells. Fortunately, under pulsed conditions
it is possible to approach the short circuit current of
the cell or batteries, about 350 mA for an 85 mAh cell
giving a pulse power of 1.96 watts from a 4 cell mercury

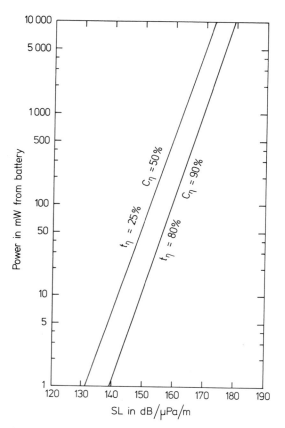

Fig. 5. Circuit parameters and battery power for acoustic power (source level).

battery. Fig. 5 is drawn from the equation:

$$SL = 170.8 + 10 \log Pe.\, t\eta.c\eta$$

The lines represent the minimum likely efficiency of the transducer and circuit, 25 and 50% maximum to 80 and 90% respectively so any normal design will achieve a source level somewhere between these for the battery power shown. Having set the source level and knowing the power required to achieve it the duration of battery life can be ascertained.

The number of days can be estimated from the expression:

$$Days = \frac{mAh}{24 Ip} \cdot \frac{10^3}{(prf)tp} \cdot \frac{24 Io}{mAh}$$

where: mAh = battery capacity in milliampere hours

p.r.f. = pulse recurrence frequency (number of pulses second^{-1})
tp = length of pulse in milliseconds
Ip = peak current drawn from battery during the pulse
Io = receiving standing current for a transponder or the 'off' current for a pinger.

The first term describes the life of the battery if a continuous wave of peak current necessary to achieve the acoustic source level were to be transmitted. This is modified by the second term for the number of times a pulse occurs in one second and its duration in milliseconds. It is this term that allows most scope for adjusting the operating life of the tag. The third term gives an approximate correction for the current taken continuously by the receiving amplifier of a transponder or the current drain during the time that a pinger is not transmitting.

Detection of Signals

Transmission rate corresponds to the amount of information to be telemetered, although it is more often used to improve the detection threshold in acoustic tag applications, i.e. a high rate will be more easily detected whether by audio or visual means. A low transmission rate can be used in practice to sample the position of the fish, for they rarely sustain speeds of over 2 body lengths sec^{-1} and it is unlikely that many hydrophone beam angles would be small enough to resolve movements occuring at intervals of less than several seconds. When the human ear is used for detection the transmission rate and pulse length relate to the recognition differential. This is defined as the amount by which the signal level exceeds the noise level presented to the ear when there is a 50% probability of detection of the signal.

The greater the bandwidth of a receiver the more noise it will let in, but for optimal detection the bandwidth must be matched to the received pulse.

$$\text{Bandwidth} \approx \frac{1}{\tau}$$

where: bandwidth is in kHz

τ = pulse length in milliseconds

Matching of this type does not preserve the pulse shape but this is of little consequence in the present application.

Acoustic Tag Construction

Being potentially disposable devices, it is important that the cost of acoustic tags is kept low. One way of achieving this is to keep the circuits as simple as possible but this limits the amount and quality of information that can be derived from biotelemetry. The smaller the

tag the less likely it is to interfere with the natural
behaviour of the fish, but it will be more difficult to
assemble by conventional techniques. Problems in the design of small acoustic tags were surveyed by Mitson and
Young (1975).

Integrated circuits have appeared in a wide range of
consumer appliances and appear to have much to offer in
terms of high packing density of components and low unit
cost. The latter is only achieved when tens of thousands
are made so that, even if the world's biotelemetry groups
each used 500 identical circuits a year, economic development would still be marginal. Although very simple circuits can be made with modest equipment, all the indications are that there will be a requirement for increasing
complexity in acoustic tags and packaged integrated
circuits will not be suitable.

Electronic components which are electrically similar
can differ greatly in physical size and this has led to
various degrees of miniaturization in the tag application
to date. In general the larger the component the easier
the construction and the lower is the labour cost. Very
little information has been published on construction,
but it generally appears to be adapted from standard
techniques. Discrete, packaged components such as transistors, resistors and capacitors are normally soldered
together, and usable assemblies have been made by doing
this without a supporting board. Apart from the high
degree of skill needed to pack the components together,
the lack of mechanical stability may cause early failure.
When an assembly of this type is encapsulated in resin
the strain on the components and joints can be severe
during the setting process.

An arrangement which allows a very high packing density
of components makes use of micro-miniature (μ min) packaged
transistors and resistors but unpackaged capacitors
(chips). These can be soldered to a flexible printed circuit board as shown in plate 2, Mitson and Young (1975),
with the advantage that it can be rolled into a cylindrical shape or folded around a rectangular component. If
made in a flat rigid form this board would be 21 x 11 mm,
too large for most tagging requirements. However, the
flexible board contains 32 of the 38 circuit elements for
a transponder and has about 7 mm^2 for each component.
This represents about the highest packing density to be
obtained by use of solder and printed circuit board assembly. The same transponder circuit has been reduced to a
circuit board 6 x 8 mm, an area reduction of 4.8 times,
with a mean area of 1.5 mm^2 for each component: no significant weight reduction is evident. To achieve the smaller
area a wire bonding technique has to be employed which
requires special facilities.

However, as the requirement increases for smaller and

more complex acoustic tags this is a likely technique. It
has the advantage that modules can be produced for the
various parts of a circuit. With proper design a transmitter circuit module would be independent of frequency and
could be coupled to a mono-stable module to make a pinger.
By attaching a receiving amplifier to the transmitter
module a transponder could be produced. In either case
the frequency conscious components of the tag, the trans-
former and transducer, would be separate anyway.

Packaging of the complete tag circuits, transducer and
battery is mainly by encapsulation in a hard setting resin
which prevents the tags being used again if recovered.
This is probably justified in the case of simple pinger
circuits but may not be so when more elaborate designs
are used. The most important factor is the minimization
of drag in external tags and a detailed analysis for a
particular design has been carried out by Arnold and
Holford (1977). This indicates that there is a need to
consider streamlining of the ends of even quite small
tags. It is also necessary to ensure that the transducer
is neither unduly damped by the material surrounding it,
nor its radiation pattern upset by adjacent components.
Activation of the tag circuits prior to use is a problem
which has attracted a number of solutions. The magnetic
reed switch can be held open by a small magnet attached
to the tag case, but the switches are bulky, not completely
reliable, and can be operated accidentally. If the tag is
large, a water activated switch could be used in seawater
applications but this is clearly not a universal solution.
Lawson and Carey (1972) designed an electronic switch
which closed once the tag was immersed in water. In most
tags wires are brought out and must be soldered together
to switch the circuit on: the only problem is that care
has to be taken to avoid water creeping down the insula-
tion and entering the circuit. Another method is to fit
two small metal pins close together so that they can be
bridged by solder to switch on.

Sensors for Biotelemetry

Much ingenuity has been shown in the design of sensors
for attachment to fish but because of the difficulty in
miniaturization the application has mainly been with the
larger species. Depth has already been mentioned because
of its importance in position fixing. Other variables
requiring fish-borne sensors are:
    heart-rate or complete ECG
    direction of movement
    speed of movement
    temperature of water surrounding the fish
    level of natural light reaching the fish.
Others such as jaw movements, tail beat, salinity of
water and the speed and direction of water currents monit-

ored from the fish may assume greater importance as techniques develop.

Kanwisher *et al.*, (1974), Wardle and Kanwisher (1974)
and Priede and Young (1977) have described experiments
to determine heart-rate and ECG. Implanted electrodes are
usually insulated stainless steel wire which has been
bared near the tip. The other electrode has been formed
by the metal housing of the tag in one case and by a tag
support wire through the base of the dorsal fin in the
other.

Direction of movement is difficult to sense but Ferrel
et al., (1974) have reported successful designs for large
sharks. A suitable design for attachment to small fish is
still awaited.

Speed of movement can be obtained by frequent measurements of range from a known point if good bearing information is also available. Ferrel *et al.*, (1974) provide
details of two sensors to measure the speed of a shark;
both rely on the drag of the device increasing as speed
increases and are unlikely to be suitable for small fish.

Temperature is an easy measurement because very small
thermistors are available which require simple circuity
to give accurate results. The simplest temperature monitoring pinger has a timing resistor replaced by a thermistor
in a manner described by Mackay (1970). In this case the
non-linear temperature resistance characteristic of thermistors will give lower accuracy at the high temperature
end of the operating range.

Light monitoring cells of high sensitivity which change
resistance and are of small size can be fitted in a similar manner to the thermistor mentioned above. It is unlikely that light alone would be monitored without depth,
so a multi-channel device might be called for, perhaps
with sequential sampling of a number of variables.

Discussion

There are some species of adult fish and the juveniles
of others, too small to be monitored by the present size
of acoustic tags. The factors that control size are:
        transducer
        battery
        complexity of circuit, i.e. number of components.
Transducer size is linked closely with frequency as shown
in Fig. 4. If high frequencies are used care must be taken
to ensure that the transducer and circuit work as efficiently as possible to give maximum acoustic power from a given
power input (Fig. 5). Above 100 kHz noise is independent
of weather conditions.

Batteries of high capacity for small volume and low
weight have a large potential market and current develop-
ments of lithium organic cells should fit the needs of the
next generation of acoustic tags. As the requirements of

biotelemetry become more exacting higher levels of electronic circuit technology can be introduced to keep size down for a given complexity. There is a limit to the amount of useful information to be obtained from a complex tag alone. At some stage it will be necessary to consider the rest of the system whether sonar or hydrophones, particularly for position fixing accuracy and its relation to fish behaviour and physiology.

References

Arnold, G.P. and Holford, B.H. (1977). *J. Cons. int. Explor. Mer*, (In press).

Ferrel, D.W., Nelson, D.R., Sciarrotta, T.C., Standora, E.A. and Carter, H.C. (1974). *Trans. Inst. Soc. Amer.*, **13**, (2) 120-131.

Hawkins, A.D., MacLennan, D.N., Urquhart, G.G. and Robb, C. (1974). *J. Fish. Biol.*, **6**, 225-236.

Harden Jones, F.R., Margetts, A.R., Greer Walker, M.G. and Arnold, G.P. (1977). *Rapp. P.-v. Reun. Cons. int. Explor. Mer*, **170**, 45-51.

Kanwisher, J., Lawson, K. and Sundnes, G. (1974). *WHOI, Fish. Bull.*,, **72**, (2) 251-253.

Lawson, K.D. and Carey, F.G. (1972). *Woods Hole Oceanographic Inst., Woods Hole, Mass., Tech. Rep.*, **71-77**, 21pp.

Luke, D.McG., Pincock, D.G. and Stasko, A.B. (1973). *J. Fish. Res. Bd Can.*, **30** (9) 1402-1404.

Mackay, R.S. (1970). *Bio-medical telemetry; sensing and transmitting biological information from animals and man*, 2nd Ed. Wiley, 533pp.

Mitson, R.B. and Cook, J.C. (1971). *Radio. Electron. Engr.*, **41**, 339-350.

Mitson, R.B. and Storeton West, T.J. (1971). *Radio Electron. Engr.*, **41**, 483-489.

Mitson, R.B. and Young, A.H. (1975). *In: Proc. Instrumentation in Oceanography*, Brit. IRE, No. 32.

Priede, I.G. and Young, A.H. (1977). *J. Fish Biol.*, **10**, 299-318.

Stasko, A.B. (1971). *Ann. N.Y. Acad. Sci.*, **188**, 12-29.

Trefethen, P.S. (1956). *Spec. Sci. Rep., Fish Wildl. Ser. US*, **179**, 11pp.

Voglis, G.M. and Cook, J.C. (1966). *Ultrasonics*, **4**, 1-9.

Wardle, C.S. and Kanwisher, J. (1974). *Mar. Behav. Physiol.*, **2**, 311-324.

Young, A.H., Tytler, P., Holliday, F.G.T. and MacFarlane, A. (1972). *J. Fish. Biol.*, **4**, 57-65.

SUMMARY OF SYMPOSIUM ON "RHYTHMIC ACTIVITY IN FISH"

J.H.S. BLAXTER

*Dunstaffnage,
Marine Research Laboratory,
Oban, Scotland.*

Introduction

This brief summary makes no attempt to catalogue the
large number of papers given at the meeting, but is rather
a comment on some of the problems still outstanding in
this field of research. I am grateful to Dr. T. Simpson
for providing me with some notes on the first day's session on the hormonal basis of rhythmic activity.

We find first that the definition of rhythms has been
quite generous - from seasonal changes in growth and reproduction, through migrations, to lunar, tidal and circadian rhythms. I suppose one could go further still
'inside' the organism and look for internal rhythms, as
indeed some workers interested in heart rate or electric
pulses did, but we are lucky that the line was drawn above
mitotic activity, and the Krebs cycle!

Hormonal Bases

The hormonal basis of rhythms was considered on the
first day. The role of the pituitary gland in fish physiology has been clouded by many earlier reports on pituitary hormones of mammalian origin, and there is a need for
more work isolating purified fish pituitary hormones,
especially for studying seasonal phenomena. Dr. Simpson
thinks we should be able to look forward to 'banks' of
hormones to which the fish physiologists might have access.

The involvement of the pineal gland in the modulation,
if not the initiation, of reproductive and pigmentation
cycles has been commented on by some participants. The
pineal may then be an area for future study.

There was some pessimism expressed on the problems of
establishing direct causal relationships between endocrine
events and changes in behaviour. This difficulty arises
from the complexity of the endocrine system and the way
in which it is modified by environmental conditions like

temperature and light.

Zeitgebers

Turning now to the so-called *zeitgeber* - this seems a useful bit of jargon but it seems clear from the meeting that we know very little about it. For example:
> What stimuli act as *zeitgebers*?
> Do two or more interact in some circumstances?
> With light what is the threshold intensity for a *zeitgeber* to operate?
> What is the action spectrum?

In fact these questions merely confirm that we know nothing about the quantitative aspects of light in animal photoperiodism generally.

Sensory Systems

Very little experimental work has been done on the sensitivity of fish to the stimuli which might constitute a *zeitgeber*. Few attempts have been made to interfere with the sensory system, most experimentation involving alteration of the stimulus itself. That in itself has caused some difficulty. Some workers use constant complete darkness (DD in the jargon); others have used a constant low light intensity often subjectively selected to suit the dark adapted performance of the human observer! It is perhaps more realistic to use low light rather than absolute darkness which only exists in caves, in burrows or in the deep sea.

Endogenous Rhythms

I am far from convinced of the value to the fish of an endogenous rhythm unless the fish requires to be *pre*-adapted to some change in the environment. If the *zeit-geber* is always present why have endogenous rhythms developed? It may be desirable for a fish to respond to the onset of a stimulus like light if it is removed from it, say in a cave or burrow, but do fish really have to be pre-adapted to awake *before* dawn and so escape their predators? One might have thought the predators would have evolved the same mechanism themselves.

Take a rhythm based on tidal transport. I can see no need for this to be endogenous since the changes caused by the tides are always present; pressure even affects animals buried in the sea bed.

This brings us to the question of *sleep* which has been used glibly and over-frequently at the meeting. What do we mean by sleep in fish? Should it not describe a true loss of awareness and the need for arousal rather than for reduced activity? Schools seem to break up at night, but what is night? Reports of intact schools at night are rarely accompanied by a light intensity measurement or the definition of a school, whether the fish are merely

aggregated or truly polarised. If schools are said to
keep together visually we need to know the visual conditions.

The Function of the Rhythms

In the long-term, rhythms are adapted to changing
seasons and probably optimum conditions for rearing the
young. In the short term how often do rhythms merely reflect sensory performance? In other words fish may not be
able to see at night and so they become inactive. This
will be an energy-sparing response.

Perhaps other rhythms which are not so easily related
to sensory performance can be seen as advantageous. This
is clear with tidal transport which may keep organisms
at an optimum depth or prevent stranding. A feeding rhythm
may be useful in preventing overfeeding. Obviously there
is no point in animals being active in food-searching if
food is not available for one reason or another. But inactive animals must beware of predators.

Techniques

We have heard described a range of techniques from the
elegant and expensive sector scanner on a research vessel
to the stopwatch in the observer's hand. Sector-scanning
probably costs £3000 per day to run, having spent a mil-
lion or so on the vessel and its equipment. It has the
advantage of working well for long periods in the sea,
perhaps in bad weather; it can show the depth of the fish
and change of position. It is difficult to say whether
the fish is swimming or exactly how it is orientated but
some inferences must be drawn from measurements of tidal
currents. A telemetering compass fixed to a tag is a likely
new development.

The stopwatch technique is cheap and has flexibility,
in that the observer can adapt to his experiment. It can
be labour-intensive if the performance of fish is to be
monitored for a long time. There are also human errors in
recording time and a limit to the information which can
be recorded in unit time by a human observer.

In between these extremes of techniques we have ultrasonic tracking, the main growth area at the moment, which
is relatively cheap in its simplest form. On-line computer
backing for position plotting will greatly increase the
expense, but without such a facility a team of workers
may be needed for a long-term study.

The sophistications of the ultrasonic tag which telemeter heartrate, tail beat or jaw movements have limitations. They need calibration. What does a particular
signal mean? ECGs are non-specific and the information
to be gleaned of different stress situations is probably
minimal.

We are also still uncertain whether tags, especially

stomach tags, affect the behaviour of the fish. Position-fixing may be also rather limited.

Radio tags with accompanying light aircraft are also being evaluated. Searching seems easier but position fixing is not as good as with acoustic tags. A light aircraft probably costs £20 per hour, 1/6 of the cost of a research boat.

The intermediate techniques, which are sometimes called actographs, and which are recording systems using light beams, thermistors, magnetic fields and so on are generally cheap, run for long periods but also require calibration. What is meant by a spike on a chart? Can one show on an actograph the difference between feeding and searching or escape from prey?

In general the laboratory experiments have lacked realism - the fish are (and must be) confined; feeding, predation, substrata are often arbitrary or not included; changes of light may not simulate dusk and dawn. But the fish is more at the behest of the investigator which is not true of ultrasonic tags; so the experimenter must choose his techniques to give the best possible data within the limitations of his methods. It is a pity that some speakers here have not been able to link experimental results with those in the wild in a meaningful way. I think there is a good case for attempting to describe, calibrate or quantify activity rhythms in some different way, perhaps by using respiration rates, total daily respiration or some factor which expresses the aggregated responses of the organism. One of the final speakers was interested in this approach.

These are just a few general remarks which reflect concern as to how far experiments can explain natural behaviour. Some investigations described during the meeting, for example fish counting, are not open to this criticism but the confines of the aquarium or the irritation of a tag may well influence activity and it is difficult to see how such possible effects can be checked.

SHORT PAPERS ALSO READ AT THE SYMPOSIUM

ADELMAN, I.R. Influence of temperature on growth promotion and body composition of carp, *Cyprinus carpio*, due to bovine growth hormone.

KATZ, A.H. and KATZ, H.M. Effects of DL-thyroxine on swimming speed in Pearl Danio, *Brachydanio albolineatus* (Blyth)

OSBORN, R.H. and SIMPSON, T.H. Seasonal changes in thyroidal status in the plaice, *Pleuronectes platessa* L.

OSBORN, R.H., SIMPSON, T.H. and YOUNGSON, A.F. Seasonal and diurnal rhythms of thyroidal status in the rainbow trout, *Salmo gairdneri* Richardson.

BYE, V.J. Annual cycles in the metabolism of the cod pineal organ.

HTUN-HAN, M. Gonad, liver and condition rhythms in dab.

HTUN-HAN, M. and BYE, V.J. Photoperiod manipulations of maturation in dab and turbot.

WHITEHEAD, C.J., BROMAGE, N.R. and FORSTER, J. Seasonal changes in reproductive function of the rainbow trout, *Salmo gairdneri*.

ATACK, T., MATTY, A.J. and SMITH, P. Feeding rhythms of rainbow trout in recirculation systems.

LANZING, W.J.R. Observations on diurnal colour pattern changes in fish.

KATZ, H.M. Circadian rhythms in juvenile American Shad, *Alosa sapidissima*.

KELSO, J.R.M. Diel rhythm in activity of Walleye, *Stizostedion vitreum vitreum*, in a small stratified temperate lake.

MACPHEE, G.K. Activity cycles of juvenile summer flounder, *Paralichthys dentatus (L).*, raised in the laboratory.

LANGFORD, T.E., FLEMING, J.M. and JAMES, N.P. A simple semi-automatic system for monitoring the migration rates of sonic-tagged salmonids in rivers.

ALABASTER, J.S. and STOTT, B. Notes on swimming activity of perch, *Perca fluviatilis*.

CRAIG, J. Diurnal rhythms of perch activity in Windermere.

COSTA, H.H. The food and feeding chronology of yellow perch, *Perca flavescens*, in Lake Washington.

LAMING, P.R. and SAVAGE, G.E. Seasonal changes in responsiveness shown by the goldfish, *Carassius auratus*.

PANNELLA, G. Potential application of otolith growth patterns to studies of fish rhythmic activity.

OTTAWAY, E. Rhythmic growth activity in fish scales.

HAWKINS, A.D. Periodicities in the feeding of juvenile cod, and in the behaviour of their prey.

HELFMAN, G. Twilight ecology of the yellow perch, *Perca flavescens*.

SJÖBERG, K. Fish activity rhythms and food-seeking behaviour of mergansers.

BRUTON, M.N. The role of diel inshore movements by *Clarias gariepinus* (Pisces: Clariidae) for the capture of fish prey.

THORPE, J.E. and MORGAN, R.I.G. Periodicity in salmon smolt migration.

McCLEAVE, J.D. Rhythmic aspects of estuarine migration of hatchery-reared Atlantic Salmon smolts, *Salmo salar*.

SOLOMON, D.J. Some observation on salmon smolt migration in a chalk stream.

TYTLER, P., THORPE, J.E. and SHEARER, W.M. Ultrasonic tracking of the movements of Atlantic salmon smolts, *Salmo salar* L., in the estuaries of two Scottish rivers.

BEACH, M.H. Activity recording using infra-red light.

GREEN, J.M. and ROYLE, J.M. A stationary underwater acoustical tracking system for fish behaviour studies in Logy Bay, Newfoundland.

OSWALD, R.L. Feeding and gill ventilation rates, and light synchronisation in wild Brown Trout, *Salmo trutta* L., revealed by ultrasonic telemetry.

MILLER, W.F. Monitoring water quality using cardiac and ventilatory activity of fish.

MESSIEH, S.N. On the regularity of spawning time of some marine fishes in the Gulf of St Lawrence.

McCLEAVE, J.D., POWER, J.H. and ROMMEL, S.A. Use of radio telemetry for studying upriver migration of adult Atlantic Salmon, *Salmo salar*.

PINCOCK, D.G. and LUKE, D. Longterm telemetry of activity information from free-swimming fish.

BERCY, C. and PIERRON, B. An automatic system for localisation of fish: operation and results.

SUBJECT INDEX

294   SUBJECT INDEX

clam, 134
clingfish *see Tomicodon*
 humeralis, Gobiesox
 pinniger
cloud cover, 110, 119, 202
cloudy weather, 221
Clupea harengus, 42, 60,
 161, 203
 maris albi, 217, 218, 221
Clupea pallasii, 203
clupeids, 203, 219, 220
coastal waters, 217, 218,
 219, 220
Cobitidae, 220
cod *see Gadus morhua*
collimation tubes, 107
Colombia, 196, 197, 239
Colossoma bidens, 194
colour temperature, 108
communication,
 acoustic, 233
 chemical, 233
 electro-, 236
competition, intraspecific,
 243, 250, 252
components, electronic, 280
Conchapelopia pellidula, 11
conductance, 236
conidia, 6
constant environmental conditions, *see also*
  Darkness, constant,
   light, constant
   209, 211
contact, intraspecific, 249
container, food-free, 244
contrast sensitivity hypothesis, 127-128
copulation, 42
Coregonus lavaretus, 14, 62,
 92, 93, 99, 225-233
corticosteroids, *see also*
 individual hormones, 21,
  58-60, 63
corticosterone, 58, 60
corticotrophin, 25, 27, 58,
 59
cortisol, 38, 39, 58, 59, 60,
 61, 62, 63, 64
cortisone, 58, 62
Coryphoblennius galerita,
 208, 209

cosine collector, 107, 109, 110,
 111, 112, 117
Cottidae, 207, 243
Cottus gobio, 70, 72, 73, 88,
 92, 96, 97, 98, 99, 243
Cottus poecilopus, 12, 13, 19,
 70, 72, 73, 88, 91, 92, 93,
 96, 97, 100
Couesius plumbeus, 38
coughs, 163
count, 254, 257, 262
counter,
 Bendix, 253-255
 conductivity, 255, 258, 259-268
 Doppler, 255-257
 electrical, 259-268
 electromechanical, 78, 226,
  255
 interrupt, 255, 257
 resistivity, 259-268
 strip, 259, 262-267
 strip, errors, 266-267
 strip, limitations, 265-266
 tube, 259-262, 263
 tube, errors, 261-262
 tube, limitations, 261
counters,
 efficiency of, 262, 267
 electronic, 263
 fish, 253-258
 siting of, 256, 258
counting fish, 253-258, 288
cover, 132, 207
crabs, 210
creeks, tidal, 207, 208
crepuscular organisms, 15, 19,
 69, 77, 98, 245
Crustacea, 194, 210, 244
curimata *see Prochilodus* spp.
current, *see also* Water velocity,
 219, 220, 221, 222, 270
current reversal, 192
curtains, acoustic, 254
cycles,
 gonadal, 39
 light-dark, 2-19, 25, 75, 76,
  79, 83, 84, 86, 91, 96, 97,
  100, 101, 132, 170, 171,
   172, 173, 174, 175, 176,
    177, 178, 179, 180, 181,
     182, 183, 199, 208,
      209, 210, 225, 232,
       236, 238